Universitext

Universitext

Universitext is a series of textbooks that presents material from a wide variety of mathematical disciplines at master's level and beyond. The books, often well class-tested by their author, may have an informal, personal even experimental approach to their subject matter. Some of the most successful and established books in the series have evolved through several editions, always following the evolution of teaching curricula, to very polished texts.

Thus as research topics trickle down into graduate-level teaching, first textbooks written for new, cutting-edge courses may make their way into *Universitext*.

For further volumes:
http://www.springer.com/series/223

Sergei Ovchinnikov

Measure, Integral, Derivative

A Course on Lebesgue's Theory

Springer

Sergei Ovchinnikov
Department of Mathematics
San Francisco State University
San Francisco, CA, USA

ISSN 0172-5939   ISSN 2191-6675 (electronic)
ISBN 978-1-4614-7195-0   ISBN 978-1-4614-7196-7 (eBook)
DOI 10.1007/978-1-4614-7196-7
Springer New York Heidelberg Dordrecht London

Library of Congress Control Number: 2013933999

Mathematics Subject Classification: 26A42, 26A45, 26A46, 28A12, 28A20, 28A25

Printed on acid-free paper

Springer is part of Springer Science+Business Media (www.springer.com)

To the memory of my best
and truest friend, Timosha

Preface

This book originated from the notes I kept while teaching the graduate course *Analysis* on Lebesgue's theory of integration and differentiation at San Francisco State University (SFSU). This course was rightly considered by many students as a difficult one, mainly because some ideas and proofs were presented in their textbooks in unnatural and counterintuitive ways, albeit rigorous ones. These students also had problems connecting the material they learned in undergraduate real analysis classes with this course. The Mathematics Department of SFSU is a master's mathematics department. Most students who received a MS degree from our department do not pursue a higher degree. When teaching this course I wanted my students to get a feel for the theory, appreciate its importance, and be ready to learn more about it, should the need arise. Consequently, my goal in writing this book was to present Lebesgue's theory in the most elementary way possible by sacrificing the generality of the theory. For this, the theory is built constructively for measures and integrals over bounded sets only. However, the reader will find all main theorems of the theory here, of course not in their ultimate generality.

The first chapter presents selected topics from the real analysis that I felt are needed to review in order to fill the gaps between what the reader probably learned some time ago or missed completely and what is required to master the material presented in the rest of the book. For instance, one can hardly find properties of summable families (Sect. 1.4) in textbooks on real analysis. Several conventions that are used throughout the book are also found in Chap. 1.

The Lebesgue measure of a bounded set and measurable functions are the subject of the second chapter. Because bounded open and closed sets have relatively simple structures, their measures are introduced first. Then the outer and inner measures of a bounded set are introduced by approximating the set by open and closed sets, respectively. A measurable set is defined as a bounded set for which its inner and outer measures are equal; its Lebesgue measure is the common value of these two measures. We proceed then by establishing standard properties of the Lebesgue measure and measurable sets. Lebesgue measurable functions and their convergence properties are covered in the last two sections of Chap. 2. Undoubtedly, the highest point of this chapter is Egorov's Theorem, which is important in establishing convergence properties of integrals in Chap. 3.

I follow most expositions in Chap. 3 where main elements of the theory of Lebesgue integral are presented. Again the theory is developed for functions over bounded sets only. However, the main convergence theorems—the Bounded Convergence Theorem, the Monotone Convergence Theorem, and Dominated Convergence Theorem—are proved in this chapter, establishing the "passage of the limit under the integral sign."

The main topics of Chap. 4 are Lebesgue's theorem about differentiability of monotone functions and his versions of the fundamental theorems of calculus. I chose to present functions of bounded variations (BV-functions) and their properties first and then prove the Lebesgue theorem for BV-functions. The proof is elementary albeit a nontrivial one. To make it more accessible, I dissect the proof into a number of lemmas and two theorems. The last two sections of Chap. 4 cover absolutely continuous functions and the fundamental theorems of calculus due to Lebesgue.

A distinguished feature of this book is that it limits attention in Chaps. 2 and 3 to bounded subsets of the real line. In the Appendix, I present a way to remedy this limitation.

There are 187 exercises in the book (there is an exercise section at the end of each chapter). Most exercises are "proof" problems, that is, the reader is invited to prove a statement in the exercise.

I have received help from many people in the process of working on the drafts of this book. First and foremost, I am greatly indebted to my students for correcting several errors in the lecture notes from which this text was derived and providing other valuable feedback. I wish to thank my colleague Eric Hayashi and an anonymous referee for reading parts of the manuscript carefully and suggesting many mathematical and stylistic corrections. My special thanks go to Sheldon Axler for his endorsement of this project and many comments which materially improved the original draft of the book. Last but not least, I wish to thank my Springer editor Kaitlin Leach for her support throughout the preparation of this book.

Berkeley, CA, USA   Sergei Ovchinnikov

Contents

1

Preliminaries

Real analysis is a standard prerequisite for a course on Lebesgue's theories of measure, integration, and derivative. The goal of this chapter is to bring readers with different backgrounds in real analysis to a common starting point. In no way the material here is a substitute for a systematic course in real analysis. Our intention is to fill the gaps between what some readers may have learned before and what is required to fully understand the material presented in the consequent chapters.

1.1 Sets and Functions

We write $x \in A$ to denote the membership of an element x in a set A. If x does not belong to the set A, then we write $x \notin A$. Two sets A and B are equal, $A = B$, if they contain the same elements, that is,

$$x \in A \quad \text{if and only if} \quad x \in B, \quad \text{for all } x.$$

A set B is a *subset* of a set A, denoted by $A \subseteq B$ (equivalently, by $B \supseteq A$), if

$$x \in B \quad \text{implies} \quad x \in A, \quad \text{for all } x.$$

Braces are frequently used to describe sets, so

$$\{x : \text{statement about } x\}$$

denotes the set of all elements x for which the statement is true. For instance, the two element set $\{1, 2\}$ can be also described as

$$\{x \in \mathbb{R} : x^2 - 3x + 2 = 0\}.$$

The operations of *intersection*, *union*, and (relative) *complement* are defined by

S. Ovchinnikov, *Measure, Integral, Derivative: A Course on Lebesgue's Theory*,
Universitext, DOI 10.1007/978-1-4614-7196-7_1,
© Springer Science+Business Media New York 2013

$$A \cap B = \{x : x \in A \quad \text{and} \quad x \in B\},$$
$$A \cup B = \{x : x \in A \quad \text{or} \quad x \in B\},$$
$$\mathsf{C}_A B = A \setminus B = \{x : x \in A \quad \text{and} \quad x \notin B\},$$

respectively, where $A \setminus B$ is the *difference* between sets A and B.

There is a unique set $\varnothing$, the *empty set*, such that $x \notin \varnothing$ for any element x. The empty set is a subset of any set. A set consisting of a single element is called a *singleton*.

The *Cartesian product* $A \times B$ of two sets A and B is the set of all ordered pairs (a, b) where $a \in A$ and $b \in B$. Two ordered pairs (a, b) and (a', b') are equal if and only if $a' = a$ and $b' = b$.

For two sets A and B, a subset $f \subseteq A \times B$ is said to be a *function* from A to B if for any element $a \in A$ there is a unique element $b \in B$ such that $(a, b) \in f$. We frequently write $b = f(a)$ if $(a, b) \in f$ and use the notation $f : A \to B$ for the function f. The sets A and B are called the *domain* and *codomain* of the function f, respectively. For a subset $A' \subseteq A$ the set

$$f(A') = \{b \in B : b = f(a), \text{ for some } a \in A'\}$$

is the *image* of A' under f. The set $f(A)$ is called the *range* of the function f. The *inverse image* $f^{-1}(B')$ of a subset $B' \subseteq B$ under f is defined by

$$f^{-1}(B') = \{a \in A : f(a) \in B'\}.$$

If $f(A) = B$, the function f is said to be *onto*. If for each $b \in f(A)$ there is exactly one $a \in A$ such that $b = f(a)$, the function f is said to be *one-to-one*. A function $f : A \to B$ is called a *bijection* if it is one-to-one and onto. In this case, we also say that f establishes a *one-to-one correspondence* between sets A and B. Given a bijection $f : A \to B$, for each element $b \in B$ there is a unique element $a \in A$ for which $f(a) = b$. Thus the function

$$f^{-1} = \{(b, a) \in B \times A : b = f(a)\}$$

is well defined. We call this function, $f^{-1} : B \to A$, the *inverse* of f.

If A and B are sets of real numbers, then a function $f : A \to B$ is called a *real function*. Real functions are the main object of study in real analysis.

For given sets A and J, a *family* $\{a_i\}_{i \in J}$ of elements of A indexed by the set J (the *index set*) is a function $a : J \to A$, that is, $a_i = a(i)$ for $i \in J$. The set $\{a_i : i \in J\}$ is the range of the function a. For $\mathcal{F} = \{a_i\}_{i \in J}$ and $a \in A$, we write $a \in \mathcal{F}$ if $a = a_i$ for some $i \in J$. If J' is a subset of the index set J, then the family $\{a_i\}_{i \in J'}$ is called a *subfamily* of the family $\{a_i\}_{i \in J}$.

If the index set J is the set of natural numbers $\mathbb{N} = \{1, 2, \ldots\}$, a family $\{a_n\}_{n \in \mathbb{N}}$ is called a *sequence* of elements of the set A. It is customary to denote a sequence by (a_n) or write it as

$$a_1, a_2, \ldots, a_n, \ldots$$

The element a_n corresponding to the index n is called the nth *term* of the sequence. A sequence is an instance of a family. However, the former has a distinguished feature—its index set $\mathbb{N}$ is an ordered set. Thus the terms of a sequence (a_n) are linearly ordered by their indices.

Let $\mathcal{F} = \{X_i\}_{i \in J}$ be a family of sets, that is, each X_i is a set. The intersection and union of $\mathcal{F}$ are defined by

$$\bigcap_{i \in J} X_i = \{x : x \in X_i \text{ for all } i \in J\}$$

and

$$\bigcup_{i \in J} X_i = \{x : x \in X_i \text{ for some } i \in J\},$$

respectively. Notations $\cap \mathcal{F}$, $\cap_{X \in \mathcal{F}} X$ and $\cup \mathcal{F}$, $\cup_{X \in \mathcal{F}} X$ are also common for these operations.

The following identities are known as *De Morgan's laws*:

$$\complement_X\left[\bigcap_{i \in J} X_i\right] = \bigcup_{i \in J}\left[\complement_X X_i\right] \quad \text{and} \quad \complement_X\left[\bigcup_{i \in J} X_i\right] = \bigcap_{i \in J}\left[\complement_X X_i\right],$$

that is, the complement of the intersection is the union of the complements, and the complement of the union is the intersection of the complements (cf. Exercises 1.2 d and 1.5 b).

A (binary) *relation* R on a set A is a subset of the Cartesian product of the set A by itself, $R \subseteq A \times A$. An *equivalence relation* on A is a relation R satisfying properties:

$(a, a) \in R$	*reflexivity*
$(a, b) \in R$ implies $(b, a) \in R$	*symmetry*
$(a, b) \in R$ and $(b, c) \in R$ imply $(a, c) \in R$	*transitivity*

for all $a, b, c \in A$. If R is an equivalence relation on A and $a \in A$, then the set

$$[a] = \{b \in A : (b, a) \in R\}$$

is called the *equivalence class* of R containing a.

Theorem 1.1. *Let R be an equivalence relation on a set A. Then:*

(i) *Any two equivalence classes of R are either identical or disjoint.*
(ii) *The set of equivalence classes partitions the set A, that is, every element of A belongs to one and only one (distinct) equivalence class.*

Proof.

(i) Suppose that $[a] \cap [b] \neq \varnothing$ and let c be an element of this intersection. Then $(c, a) \in R$ and, by symmetry, $(a, c) \in R$. For any $x \in [a]$, we have

$$(x, a) \in R, \quad (a, c) \in R, \quad \text{and} \quad (c, b) \in R.$$

By transitivity, $(x, b) \in R$, that is, $x \in [b]$. Hence, $[a] \subseteq [b]$. By reversing the roles of $[a]$ and $[b]$, we obtain $[b] \subseteq [a]$. Therefore, $[a]=[b]$ if $[a] \cap [b] \neq \varnothing$.

(ii) Every element a of A is in the equivalence class $[a]$. This equivalence class is unique by part (i). $\qquad\square$

A family of nonempty subsets $\{A_i\}_{i \in J}$ of a set A is said to be a *partition* of A if

$$\bigcup_{i \in J} A_i = A \quad \text{and} \quad A_i \cap A_j = \varnothing, \text{ for all } i \neq j \text{ in } J.$$

For a given partition $\{A_i\}_{i \in J}$ of A, we define a binary relation R on A by

$$(a, b) \in R \quad \text{if and only if} \quad a, b \in A_i, \text{ for some } i \in J.$$

It can be readily verified that the relation R is an equivalence relation on A (cf. Exercise 1.14).

1.2 Sets and Sequences of Real Numbers

Throughout the book the symbols $\mathbb{R}$ and $\mathbb{Q}$ denote the sets of real and rational numbers, respectively.

Let a and b be real numbers. A bounded *open interval* (a, b) is the set $(a, b) = \{x : a < x < b\}$. The sets $[a, b) = \{x : a \leq x < b\}$ and $(a, b] = \{x : a < x \leq b\}$ are called *half-open intervals*. For $a \leq b$, a bounded *closed interval* is the set $[a, b] = \{x : a \leq x \leq b\}$. The following sets are *unbounded intervals*:

$$(a, +\infty) = \{x : x > a\}, \quad (-\infty, a) = \{x : x < a\} \qquad \text{(open)},$$
$$[a, +\infty) = \{x : x \geq a\}, \quad (-\infty, a] = \{x : x \leq a\} \qquad \text{(closed)},$$
$$(-\infty, +\infty) = \mathbb{R} \qquad \text{(both)}.$$

An *interval* is any set defined in the foregoing paragraph. The points a and b are called the *endpoints* of the respective intervals.

If $a > b$, then all bounded intervals (a, b), $[a, b)$, $(a, b]$, and $[a, b]$ are the empty set. If $b = a$, the intervals (a, a), $[a, a)$, and $(a, a]$ are the empty set, while $[a, a] = \{a\}$. These intervals are called *degenerate*. In the book, we mostly restrict our analysis to nondegenerate intervals.

The symbols $-\infty$ and $+\infty$ are not elements of $\mathbb{R}$. Sometimes it is convenient to add these symbols to $\mathbb{R}$ and then call the resulting set $\mathbb{R} \cup \{-\infty, +\infty\}$ the *extended real numbers*. We extend the inequality relation $<$ to $\mathbb{R} \cup \{-\infty, +\infty\}$ by setting $-\infty < a$ and $a < \infty$ for any $a \in \mathbb{R}$.

A nonempty set E of real numbers is said to be *bounded above* if there is a real number b, which is called an *upper bound* for E, such that $x \leq b$ for all $x \in E$. Note that if E has an upper bound b then any b' such that $b < b'$ is also an upper bound for E. If a number c is not an upper bound for E, then there is $x \in E$ such that $x > c$ (cf. Exercise 1.16).

The following completeness property is one of the fundamental properties of the set of real numbers.

The Completeness Property. Let E be a nonempty set of real numbers that is bounded above. Then among the upper bounds for E there is a least upper bound.

It is not difficult to show that there can be only one least upper bound for a bounded set E, the existence of which is asserted by the completeness property (cf. Exercise 1.17). This least upper bound is called the *supremum* of E and denoted by $\sup E$.

Similarly, we say that a nonempty set $E \subseteq \mathbb{R}$ is *bounded below* if there is $b \in \mathbb{R}$ such that $x \geq b$ for all $x \in E$. Then the number b is called *lower bound* for E. It follows from the completeness property that a nonempty set E of real numbers that is bounded below has the greatest lower bound which is called the *infimum* of E and denoted by $\inf E$ (cf. Exercise 1.18).

The concepts introduced above are illustrated by the drawing in Fig. 1.1.

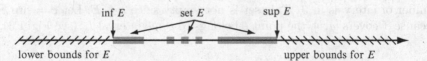

inf E set E sup E

lower bounds for E upper bounds for E

Figure 1.1. Infimum and supremum of E

A nonempty set of real numbers is said to be *bounded* if it is both bounded above and below. Otherwise, the set is called *unbounded*.

One should distinguish the supremum and the infimum of a bounded nonempty set from its maximum and minimum that are the greatest and the least numbers in the set, respectively (cf. Exercise 1.24).

A function $f : A \to \mathbb{R}$ is said to be bounded if its range $f(A)$ is bounded.

We will write $\sup E = +\infty$ for a set E that is not bounded above. Similarly, we will write $\inf E = -\infty$ if the set E is not bounded below.

Let E be a set of real numbers that is bounded above. It is clear that an upper bound u for E is not the supremum of E if and only if there is an upper bound v for E such that $v < u$, or equivalently, by letting $\varepsilon = u - v > 0$, an upper bound u for E is not the supremum of E if and only if there is $\varepsilon > 0$ such that $x \leq u - \varepsilon$ for all $x \in E$. We reformulate the last biconditional statement as follows:

Approximation Property of Supremum. An upper bound u for a set E is the supremum of E if and only if for each $\varepsilon > 0$ there exists $x \in E$ such that $x > u - \varepsilon$ (see the drawing in Fig. 1.2).

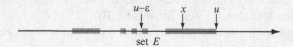

Figure 1.2. $u = \sup E$.

As an immediate application of the approximation property, we establish an important property of closed intervals in the number line.

A family $\{I_i\}_{i \in J}$ of open intervals is said to be a *cover* of a set E provided that $E \subseteq \bigcup_{i \in J} I_i$. By a *subcover* of a cover of E we mean a subfamily of the cover that itself also is a cover of E. We use the result of the following theorem, in Chap. 2.

Theorem 1.2. (Heine–Borel Theorem) *Every cover of a closed interval by open intervals contains a finite subcover.*

Proof. Let $\mathcal{F} = \{I_i\}_{i \in J}$ be a cover of $[a, b]$ by open intervals. Let us define E to be the set of points $x \in [a, b]$ such that the interval $[a, x]$ is covered by a finite number of intervals in $\mathcal{F}$. This set is not empty since $a \in E$. Let $c = \sup E$. Because $\mathcal{F}$ covers $[a, b]$, the point c belongs to some interval I_k (see Fig. 1.3).

Figure 1.3. Heine–Borel theorem

By the approximation property of supremum, there is $x \in E$ such that $x \in I_k$. Since $[a, c] \subseteq [a, x] \cup I_k$, the interval $[a, c]$ is covered by a finite subfamily of $\mathcal{F}$. If $c = b$, then we are done. Suppose that $c < b$. Then there is $y \in I_k$ such that $c < y < b$. Hence, the interval $[a, y] \subseteq [a, c] \cup I_k$ is covered by a finite subfamily of $\mathcal{F}$. This contradicts our choice of $c = \sup E$. It follows that $[a, b]$ is covered by a finite subfamily of $\mathcal{F}$. $\square$

Now we recall the definition of a convergent sequence of real numbers.

Definition 1.1. *A number a is said to be a limit of a sequence (a_n) provided that for any $\varepsilon > 0$ there exists $N \in \mathbb{N}$ such that*

$$|a - a_n| < \varepsilon, \qquad \text{for all } n \geq N$$

or equivalently

$$a - \varepsilon < a_n < a + \varepsilon, \qquad \text{for all } n \geq N.$$

Then the sequence (a_n) is said to converge *to the number a and the convergence of (a_n) is denoted by*

$$a_n \to a \qquad \text{or} \qquad \lim a_n = a.$$

Note that this definition employs the natural order on the set $\mathbb{N}$.

A sequence of real numbers (a_n) is said to be *bounded above (below)* if the set $\{a_n : n \in \mathbb{N}\}$ is bounded above (below). We will use notations $\sup_n a_n$ and $\inf_n a_n$ for the supremum and infimum of the set $\{a_n : n \in \mathbb{N}\}$, respectively.

A sequence (a_n) is said to be *increasing (decreasing)* provided that $a_n \leq a_{n+1}$ $(a_n \geq a_{n+1})$ for all n:

$$a_1 \leq a_2 \leq \cdots \leq a_n \leq a_{n+1} \leq \cdots \qquad \text{increasing sequence,}$$
$$a_1 \geq a_2 \geq \cdots \geq a_n \geq a_{n+1} \geq \cdots \qquad \text{decreasing sequence.}$$

Clearly, an increasing (decreasing) sequence is bounded below (above). A sequence is said to be *monotone* provided it is either increasing or decreasing.

Theorem 1.3. *A monotone sequence converges if and only if it is bounded. A bounded increasing sequence converges to its supremum, whereas a bounded decreasing sequence converges to its infimum.*

Proof. We consider the case when the sequence is increasing. The proof for a decreasing sequence is similar.

(Necessity.) Suppose $a_n \to a$. Then, for $\varepsilon = 1$, there is only finite number of terms a_n's that lie outside the interval $(a - 1, a + 1)$. Thus, (a_n) is bounded.

(Sufficiency.) Let $a = \sup_n a_n$. By the approximation property of supremum, for a given $\varepsilon > 0$ there is N such that $a_N > a - \varepsilon$. Because (a_n) is an increasing sequence, we have $a_n > a - \varepsilon$ for every $n \geq N$. Hence, $a - a_n < \varepsilon$ for $n \geq N$; that is, $\lim a_n = \sup_n a_n$. □

Now let (a_n) be an arbitrary bounded sequence. Let us define $b_n = \sup_{k \geq n} a_k = \sup\{a_k : k \geq n\}$ for $n \in \mathbb{N}$. Because

$$\sup_{k \geq n+1} a_k \leq \sup_{k \geq n} a_k$$

(cf. Exercise 1.25), (b_n) is a bounded decreasing sequence. By Theorem 1.3, the sequence (b_n) is convergent. Its limit is called the *limit superior* of the sequence (a_n) and denoted by $\limsup a_n$ (or by $\overline{\lim} a_n$). Thus,

$$\limsup a_n = \lim(\sup\{a_k : k \ge n\}). \tag{1.1}$$

Similarly, the sequence $c_n = \inf_{k \ge n} a_k$ is a bounded increasing sequence. Its limit is denoted by $\liminf a_n$ (or by $\underline{\lim} a_n$) and called *limit inferior* of the sequence (a_n).

Theorem 1.4. *Let (a_n) be a bounded sequence of real numbers. Then $\limsup a_n = a$ if and only if for each $\varepsilon > 0$, there are infinitely many indices n for which $a_n > a - \varepsilon$ and at most finitely many indices n for which $a_n > a + \varepsilon$ (cf. Fig. 1.4).*

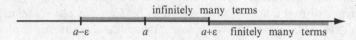

infinitely many terms

$a{-}\varepsilon$ \qquad a \qquad $a{+}\varepsilon$ \quad finitely many terms

Figure 1.4. Theorem 1.4

Proof. (Necessity.) Let $\limsup a_n = a$. Then, for a given $\varepsilon > 0$, there is N such that (cf. (1.1))

$$a - \varepsilon < \sup_{k \ge n} a_k < a + \varepsilon, \qquad \text{for all } n \ge N. \tag{1.2}$$

Suppose that there are only finitely many indices n for which $a_n > a - \varepsilon$ and let m be the greatest of these indices. Then $a_n \le a - \varepsilon$ for all $n > m$. Let n be an index greater than both m and N. Then $\sup_{k \ge n} a_k \le a - \varepsilon$, which contradicts the left inequality in (1.2). Hence there are infinitely many indices n for which $a_n > a - \varepsilon$.

By the right inequality in (1.2), $\sup_{k \ge N} a_k < a + \varepsilon$. Therefore, $a_k < a + \varepsilon$ for all $k \ge N$. It follows that there are only finitely many indices n for which $a_n > a + \varepsilon$.

(Sufficiency.) Let $\varepsilon > 0$ be a given number. There are only finitely many indices n for which $a_n > a + \varepsilon/2$, therefore there is an index N such that $a_n \le a + \varepsilon/2$ for all $n \ge N$. Hence,

$$\sup_{k \ge n} a_k \le a + \varepsilon/2 < a + \varepsilon, \qquad \text{for all } n \ge N.$$

Because there are infinitely many indices n for which $a_n > a - \varepsilon$, for any $n \ge N$ there is $k \ge n$ such that $a_k > a - \varepsilon$. Therefore,

$$a - \varepsilon < \sup_{k \ge n} a_k, \qquad \text{for all } n \ge N.$$

It follows that for every $\varepsilon > 0$ there is N such that

$$a - \varepsilon < \sup_{k \ge n} a_k < a + \varepsilon, \qquad \text{for all } n \ge N,$$

that is, $\limsup a_n = a$ (cf. (1.1)). $\qquad\qquad\qquad\qquad\qquad\qquad\qquad\square$

A similar criterion holds for the limit inferior of a bounded sequence; see Exercise 1.29.

Example 1.1. Since the set of rational numbers in the interval $[0,1]$ is countable, it can be enumerated into a sequence $r_1, r_2, \ldots, r_n \ldots$. For any given $\varepsilon > 0$, there are infinitely many r_n's greater than $1 - \varepsilon$, and none are greater than $1 + \varepsilon$. Thus, $\limsup r_n = 1$. Similarly, $\liminf r_n = 0$.

The result of the next theorem follows almost immediately from the criteria in Theorem 1.4 and Exercise 1.29 (cf. Exercise 1.32).

Theorem 1.5. *A bounded sequence (a_n) is convergent if and only if* $\liminf a_n = \limsup a_n$. *In this case,*

$$\lim a_n = \liminf a_n = \limsup a_n.$$

1.3 Open and Closed Sets of Real Numbers

Definition 1.2. *A set G of real numbers is called* open *provided that for each point $x \in G$ there is an open interval containing x which is contained in G.*

All open intervals, including $\mathbb{R}$ itself, are open sets. Since the empty set $\varnothing$ contains no points, it is vacuously open.

Theorem 1.6. *The union of an arbitrary family of open sets is open. The intersection of a finite family of open sets is open.*

Proof. The first claim is obvious. Suppose $G = \bigcap_{i=1}^{n} G_i$ is a finite intersection of open sets. If $G = \varnothing$, we are done. Suppose that $G \neq \varnothing$ and let $x \in G$. Then $x \in G_i$ for all $1 \leq i \leq n$. Because the sets G_i's are open, there are open intervals (a_i, b_i) such that $x \in (a_i, b_i)$ and $(a_i, b_i) \subseteq G_i$ for $1 \leq i \leq n$. It remains to note that the open interval (a, b) with $a = \max\{a_i : 1 \leq i \leq n\}$ and $b = \min\{a_i : 1 \leq i \leq n\}$ is a subset of $G = \bigcap_{i=1}^{n} G_i$ containing x (cf. Exercise 1.38). $\qquad\square$

Theorem 1.7. *Every nonempty open set G is the union of a finite or countable family of pairwise disjoint open intervals. These open intervals are called* component intervals *of G.*

Proof. Let x be a point in a nonempty open set G. There is an open interval (y, z) such that $x \in (y, z) \subseteq G$. Then, $(y, x) \subseteq G$ and $(x, z) \subseteq G$. We define (possibly extended) numbers a_x and b_x by

$$a_x = \inf\{y : (y, x) \subseteq G\} \quad \text{and} \quad b_x = \sup\{z : (x, z) \subseteq G\}.$$

It is clear that $x \in I_x = (a_x, b_x)$. We claim that $I_x \subseteq G$ but $a_x \notin G$, $b_x \notin G$ and prove this assertion in the next two paragraphs.

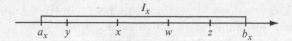

Figure 1.5. Proof of Theorem 1.7

First we prove that $I_x \subseteq G$ (cf. Fig. 1.5). Let $w \in I_x$ and assume that $w > x$ (The case when $w < x$ is treated similarly.) By the definition of b_x, there is $z > w$ such that $(x, z) \subseteq G$. Then $w \in (x, z) \subseteq G$. It follows that $I_x \subseteq G$.

Now suppose that $b_x \in G$. Then there is an interval $(u, v) \subseteq G$ with $u < b_x < v$. Clearly, $(x, b_x) \cap (u, v) \neq \varnothing$. We obtain a desired contradiction by noting that $(x, v) \subseteq G$ (cf. Fig. 1.6). A similar argument shows that $a_x \notin I_x$.

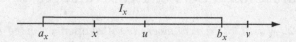

Figure 1.6. Proof of Theorem 1.7

Inasmuch as each $x \in G$ is an element of I_x and each I_x is contained in G, we have $G = \bigcup_{x \in G} I_x$. Suppose that two distinct intervals I_x and I_y intersect. Then one of the endpoints of one of these intervals belongs to the other interval—a contradiction, since endpoints of the intervals do not belong to G. Thus G is a union of a family of pairwise disjoint intervals. Because each of these intervals contains a rational number, G is a union of a finite or countable family of pairwise disjoint intervals (cf. Exercise 1.37).   □

Corollary 1.1. *If a nonempty open set G_1 is a subset of an open set G_2, then each component interval of G_1 is a subset of a unique component interval of G_2.*

Proof. Let (a, b) be a component interval of G_1 and x be a point in (a, b). By the definition of the component interval $I_x = (a_x, b_x)$ of the set G_2 (cf. the proof of Theorem 1.7), we have

$$a_x \leq a < b \leq b_x.$$

Thus $(a, b) \subseteq I_x$. Because the component intervals of G_2 are pairwise disjoint, the interval I_x is a unique component interval of G_2 containing (a, b).   □

Theorem 1.8. *If an open set G is the union of a family $\{G_i\}_{i \in J}$ of pairwise disjoint open sets, then each component interval of G is a component interval of one of the sets G_i's and the component intervals of the sets G_i's are component intervals of G.*

Proof. Because $G = \bigcup_{i \in J} G_i$ and each set G_i is the union of its component intervals, the component intervals of G are component intervals of the sets G_i's (cf. Exercise 1.39).

Let (a, b) be a component interval of some set G_k. By Corollary 1.1, the interval (a, b) is contained in some component interval (x, y) of the set G. The endpoint b does not belong to the set G, for otherwise it would belong to a component interval of a set G_i with $i \neq k$, which contradicts our assumption that the sets G_i's are pairwise disjoint. Likewise, a does not belong to G. It follows that $(a, b) = (x, y)$; that is, the component intervals of the sets G_i's are component intervals of the set G. $\square$

Definition 1.3. *A set F of real numbers is said to be* closed *if its complement* $\complement_{\mathbb{R}} F$ *is open.*

In what follows, we denote by $\complement X$ the complement of a set X in $\mathbb{R}$, so $\complement X = \complement_{\mathbb{R}} X = \mathbb{R} \setminus X$.

By De Morgan's laws

$$\complement \bigcup_{i \in J} X_i = \bigcap_{i \in J} \complement X_i \quad \text{and} \quad \complement \bigcap_{i \in J} X_i = \bigcup_{i \in J} \complement X_i.$$

Therefore, Theorem 1.6 may be reformulated in terms of closed sets as follows.

Theorem 1.9. *The empty set $\varnothing$ and $\mathbb{R}$ are closed sets. The union of a finite family of closed sets is closed. The intersection of an arbitrary family of closed sets is closed.*

We need the results of the following three theorems in Chap. 2.

Theorem 1.10. *Let F be a nonempty bounded closed set. Then points $a = \inf F$ and $b = \sup F$ belong to the set F and the set $\complement_{[a,b]} F = [a, b] \setminus F$ is open.*

Proof. Inasmuch as a is the infimum of F, every open interval containing a must intersect the set F. Hence a cannot belong to the open set $\complement F$. It follows that $a \in F$. Similar argument shows that $b \in F$. Clearly, $F \subseteq [a, b]$. Since $a, b \in F$, we have

$$[a, b] \setminus F = (a, b) \setminus F = (a, b) \cap \complement F,$$

which is an open set because the set F is closed. $\square$

Theorem 1.11. *The only subsets of $\mathbb{R}$ that are both open and closed are $\varnothing$ and $\mathbb{R}$.*

Proof. The proof is by contradiction. Suppose that X is a subset of $\mathbb{R}$ which is both open and closed and distinct from $\varnothing$ and $\mathbb{R}$. Then at least one of the component intervals of X has an endpoint $a \notin X$. Any open interval containing a intersects X. Therefore, $\complement X$ is not an open set which contradicts our assumption that X is closed. $\qquad\square$

Theorem 1.12. *Let F_1 and F_2 be two disjoint closed subsets of $\mathbb{R}$. There are open sets G_1 and G_2 such that $F_1 \subseteq G_1$, $F_2 \subseteq G_2$, and $G_1 \cap G_2 = \varnothing$.*

Proof. For $x \in F_1$, we define

$$\rho_x = \inf\{|x - y| : y \in F_2\}.$$

Because $\complement F_2$ is an open set containing x, there is $\delta > 0$ such that

$$(x - \delta, x + \delta) \subseteq \complement F_2.$$

Hence, $\rho_x > \delta > 0$. Similarly, for $y \in F_2$, we define

$$\rho_y = \inf\{|y - x| : x \in F_1\}$$

with $\rho_y > 0$.

Let open intervals I_x and I_y be defined by

$$I_x = (x - \tfrac{\rho_x}{2}, x + \tfrac{\rho_x}{2}), \quad I_y = (y - \tfrac{\rho_y}{2}, y + \tfrac{\rho_y}{2}).$$

Consider the open sets

$$G_1 = \bigcup_{x \in F_1} I_x, \quad G_2 = \bigcup_{y \in F_2} I_y.$$

Clearly, $F_1 \subseteq G_1$ and $F_2 \subseteq G_2$. Suppose that there is $z \in G_1 \cap G_2$. Then there are $x \in F_1$ and $y \in F_2$ such that $z \in I_x$ and $z \in I_y$. Hence, $|x - z| < \frac{\rho_x}{2}$ and $|y - z| < \frac{\rho_y}{2}$, and

$$|x - y| \le |x - z| + |y - z| < \tfrac{\rho_x + \rho_y}{2}.$$

By symmetry, we may assume that $\rho_x \ge \rho_y$. Then $\frac{\rho_x + \rho_y}{2} \le \rho_x$. From the previous displayed inequality we have $|x - y| < \rho_x$. This contradicts the definition of ρ_x. It follows that $G_1 \cap G_2 = \varnothing$. $\qquad\square$

Despite their simple descriptions, open and closed sets of real numbers may be quite complex, as the following example demonstrates.

Example 1.2. (The Cantor set) The Cantor set can be described by removing a sequence of open intervals from the interval $I = [0, 1]$. First, we remove the open interval $(\frac{1}{3}, \frac{2}{3})$ from I to obtain the set

$$C_1 = \left[0, \frac{1}{3}\right] \cup \left[\frac{2}{3}, 1\right]$$

(see Fig. 1.7).

Second, we remove the open middle third of each of the two closed intervals in C_1 to obtain the set

$$C_2 = \left[0, \frac{1}{9}\right] \cup \left[\frac{2}{9}, \frac{1}{3}\right] \cup \left[\frac{2}{3}, \frac{7}{9}\right] \cup \left[\frac{8}{9}, 1\right].$$

Figure 1.7. Construction of the Cantor set

Note that C_2 is the union of $4 = 2^2$ closed intervals, each of which has length $\frac{1}{9} = \frac{1}{3^2}$. By continuing in this way, we construct the set C_k which is the union of 2^k closed intervals of the form $\left[\frac{m}{3^k}, \frac{m+1}{3^k}\right]$. The next set C_{k+1} is obtained by removing the open middle third from each of these intervals. This process can be described by the recurrence equation

$$C_{k+1} = \tfrac{1}{3}C_k \cup \left(\tfrac{2}{3} + \tfrac{1}{3}C_k\right) \tag{1.3}$$

(cf. Exercise 1.44).

By definition, the *Cantor set* $\mathbf{C}$ is the intersection of the closed sets C_k:

$$\mathbf{C} = \bigcap_{k=1}^{\infty} C_k,$$

and therefore itself is a closed set. This set contains all of the endpoints of the removed open intervals. Therefore, the Cantor set contains infinitely many points. The complement $\complement_I \mathbf{C} = I \setminus \mathbf{C}$ is the union of all removed open intervals and therefore is an open set (cf. Theorem 1.10).

Note that the total length of the removed intervals is 1, since

$$\frac{1}{3} + \frac{2}{3^2} + \cdots + \frac{2^n}{3^{n+1}} + \cdots = 1.$$

Thus $\mathbf{C}$ is a subset of the unit interval I whose complement in I has "total length" 1.

Let us enumerate intervals in the sets C_k as follows. We denote I_0 and I_1 the left and the right intervals in the set C_1, respectively. The left and right subintervals of I_0 that belong to C_2 are denoted by I_{00} and I_{01}, and the left and right subintervals of I_1 are denoted by I_{10} and I_{11} (cf. Fig. 1.8). If $I_{i_1 \dots i_k}$ is an interval in the set C_k, then its left and right subintervals in the set C_{k+1} are $I_{i_1 \dots i_k 0}$ and $I_{i_1 \dots i_k 1}$, respectively. Note that by this way of denoting subintervals of the sets C_k, they are enumerated from left to right by binary numbers from 0 to $2^k - 1$.

Any number x in the Cantor set $\mathbf{C}$ belongs to one and only one of the subintervals of each set C_k, $k \in \mathbb{N}$. Therefore, elements of $\mathbf{C}$ are in one-to-one correspondence with the nested sequences of intervals

$$I_{i_1} \supset I_{i_1 i_2} \supset \cdots \supset I_{i_1 i_2 \dots i_k} \supset \cdots,$$

and therefore in one-to-one correspondence with the sequences (i_k), where $i_k \in \{0, 1\}$ for $k \in \mathbb{N}$. It follows that the cardinality of the Cantor set is $2^{\aleph_0}$, that is, the cardinality $\mathfrak{c}$ of the continuum $\mathbb{R}$.

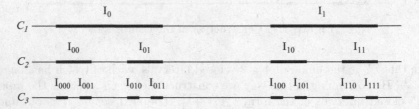

Figure 1.8. Intervals in the sets C_1–C_3

1.4 Summation on Infinite Sets

We begin by recalling the definition of a convergent series.

Definition 1.4. *If (a_n) is a sequence of real numbers and*

$$s_n = \sum_{i=1}^{n} a_i,$$

then the sequence (s_n) is called a series and the terms of (a_n) are called the terms of the series. The series (s_n) is denoted by $\sum a_i$. If the series $\sum a_i$ converges to some number S, we say that S is the sum of the series and write $\sum_{i=1}^{\infty} a_i = S$ (or $\sum_{i \in \mathbb{N}} a_i = S$, or simply $\sum_i a_i = S$).

Note that this definition utilizes the order of the set of natural numbers $\mathbb{N}$.

For an arbitrary family of real numbers $\{a_i\}_{i \in J}$ (without any assumption made about the index set J), we adopt the following definition.

Definition 1.5. *Let* $\{a_i\}_{i \in J}$ *be a family of real numbers. With each finite subset* K *of* J *we associate the number* $S_K = \sum_{i \in K} a_i$ *and call it the* finite partial sum *of the family* $\{a_i\}_{i \in J}$*, corresponding to the set* K*. The family* $\{a_i\}_{i \in J}$ *is* summable *and its* sum *is* S *if, for each* $\varepsilon > 0$*, there is a finite subset* J_0 *of* J *such that for each finite subset* $K \supseteq J_0$ *of* J *we have* $|S - S_K| < \varepsilon$*. In this case we also say that the family* $\{a_i\}_{i \in J}$ *is summable to* S *and write* $\sum_{i \in J} a_i = S$.

If a sequence (a_n) is summable to S, then the series $\sum a_i$ converges to S. Indeed, for a given $\varepsilon > 0$, let J_0 be the set from Definition 1.5 and let $N = \max J_0$. Then, by the definition of J_0, we have

$$\left| S - \sum_{i=1}^{n} a_i \right| < \varepsilon, \qquad \text{for all } n > N,$$

so $\sum_{i=1}^{\infty} a_i = S$. Note that the converse is not true; see Exercise 1.50.

Because the definition of a summable family does not involve any ordering of the index set J, we may say that the notion of a sum enjoys the *commutativity* property. More precisely, we have the following theorem.

Theorem 1.13. *Let* $\{a_i\}_{i \in J}$ *be a summable family and let* $f : J \to J$ *be a bijection (a permutation of* J*). We define* $b_i = a_{f(i)}$ *for all* $i \in J$*. Then the family* $\{b_i\}_{i \in J}$ *is summable and has the same sum as* $\{a_i\}_{i \in J}$.

Proof. Let S be the sum of $\{a_i\}_{i \in J}$. Then for each $\varepsilon > 0$ there is a finite set $J_0 \subseteq J$ such that $|S - S_K| < \varepsilon$ for any finite subset K of J containing J_0. Consider an arbitrary finite subset L of J containing the finite set $f^{-1}(J_0)$; that is, $L \supseteq f^{-1}(J_0)$. Then

$$\sum_{i \in L} b_i = \sum_{i \in L} a_{f(i)} = \sum_{f^{-1}(j) \in L} a_j = \sum_{j \in f(L)} a_j.$$

Since $f(L) \supseteq J_0$, we have $|S - \sum_{i \in L} b_i| < \varepsilon$, and the result follows.   □

In the case of a countable index set J we may assume that $J = \mathbb{N}$. Then the family $\{a_i\}_{i \in J}$ is a sequence (a_n) and Theorem 1.13 tells us that the terms of the series $\sum_{i \in \mathbb{N}} a_i$ can be arbitrary rearranged without changing summability of the series and its sum.

If the terms of a family $\{a_i\}_{i \in J}$ are nonnegative numbers, we have the following criterion for summability of the family.

Theorem 1.14. *A family of nonnegative numbers* $\{a_i\}_{i \in J}$ *is summable if and only if the set of its finite partial sums is bounded above. If so, the supremum of this set is the sum of the family* $\{a_i\}_{i \in J}$.

Proof. (Necessity.) Suppose that $\sum_{i \in J} a_i = S$. First, we show that S is an upper bound for the finite partial sums. Suppose to the contrary that there is a finite set $L \subseteq J$ such that $S_L > S$ and let $\varepsilon \doteq S_L - S$. Because the family $\{a_i\}_{i \in J}$ is summable, there is a finite set $J_0 \subseteq J$ such that, for any finite subset K of J containing J_0, we have

$$S - \varepsilon < S_K < S + \varepsilon.$$

For $K = J_0 \cup L$, we have $S_K \geq S_L$, because the numbers a_i's are nonnegative. By the definition of ε, we have $S_L - S + \varepsilon$, which contradicts the last displayed inequality. Hence, S is an upper bound for the finite partial sums.

Now, for a given $\varepsilon > 0$, let J_0 be the finite set from Definition 1.5. Then $|S - S_{J_0}| < \varepsilon$, that is, $S - \varepsilon < S_{J_0}$. By the approximation property of supremum, S is the supremum of the family of all finite partial sums.

(Sufficiency.) Let S be the supremum of the family of all finite partial sums and let ε be a positive number. By the approximation property of supremum, there is a finite set of indices J_0 such that $S - \varepsilon < S_{J_0} \leq S$. For any finite set $K \supseteq J_0$, we have $S_{J_0} \leq S_K$, for the numbers a_i are nonnegative. Therefore

$$S - \varepsilon < S_{J_0} \leq S_K \leq S,$$

so, according to Definition 1.5, the family $\{a_i\}_{i \in J}$ is summable and its sum is S.   □

Corollary 1.2. *Every subfamily of a summable family of nonnegative numbers is summable. Furthermore, the sum of the subfamily is not greater than the sum of the family.*

In fact, the result of Corollary 1.2 holds for an arbitrary summable family of real numbers. We omit the proof of this claim and, in what follows, consider only families of nonnegative numbers.

Theorem 1.15. (Principle of comparison) *Let $\{a_i\}_{i \in J}$ and $\{b_i\}_{i \in J}$ be two families of real numbers such that $0 \leq a_i \leq b_i$ for all $i \in J$. If the family $\{b_i\}_{i \in J}$ is summable then so is $\{a_i\}_{i \in J}$ and we have*

$$\sum_{i \in J} a_i \leq \sum_{i \in J} b_i.$$

Theorem 1.15 provides the most commonly used criterion for deciding whether or not a sequence of nonnegative real numbers is summable. The proof is left to the reader as Exercise 1.52.

When dealing with a finite sum of numbers, we can associate its summands arbitrarily into groups, add the numbers in each group, and then add the resulting sums to obtain the total sum. This property is known as the *associativity* property of addition. We show that any summable family of nonnegative real numbers enjoys the same property.

Let $\{a_i\}_{i \in J}$ be a summable family of nonnegative real numbers and let

$$J = \bigcup_{\lambda \in \Lambda} J_\lambda$$

be a partition of J into a family of subsets $\{J_\lambda\}_{\lambda \in \Lambda}$. Thus, we assume that $J_\lambda \neq \varnothing$ and $J_\lambda \cap J_{\lambda'} = \varnothing$ for all $\lambda \neq \lambda'$ in Λ. By Corollary 1.2, each subfamily $\{a_i\}_{i \in J_\lambda}$ is summable. Let us denote its sum by S_λ.

Theorem 1.16. *The family $\{S_\lambda\}_{\lambda \in \Lambda}$ is summable and has the same sum as the family $\{a_i\}_{i \in J}$, that is,*

$$\sum_{\lambda \in \Lambda} \Big(\sum_{i \in J_\lambda} a_i \Big) = \sum_{i \in J} a_i.$$

Proof. Let $S = \sum_{i \in J} a_i$. We need to show that for each $\varepsilon > 0$, there is a finite set $\Lambda_0 \subseteq \Lambda$ such that $|S - \sum_{\lambda \in \Phi} S_\lambda| < \varepsilon$ for any finite set $\Phi \subseteq \Lambda$ containing Λ_0.

Let ε be a given positive number. Because $\{a_i\}_{i \in J}$ is summable to S, there is a finite set $J_0 \subseteq J$ such that $|S - \sum_{i \in K} a_i| < \varepsilon/2$ for any finite set $K \supseteq J_0$. We define Λ_0 as the set of all indices λ such that $I_\lambda = J_\lambda \cap J_0 \neq \varnothing$ and show that Λ_0 is the desired set. Clearly, the set Λ_0 is finite.

Let Φ be a finite subset of Λ, containing Λ_0, and let $n = |\Phi|$ be the number of elements in Φ. By the definition of S_λ, for each λ there is a finite subset H_λ of J_λ, containing I_λ and such that

$$\Big| S_\lambda - \sum_{i \in H_\lambda} a_i \Big| < \frac{\varepsilon}{2n}$$

(cf. Exercise 1.48). We have

$$\Big| \sum_{\lambda \in \Phi} S_\lambda - \sum_{\lambda \in \Phi} \Big(\sum_{i \in H_\lambda} a_i \Big) \Big| \leq \sum_{\lambda \in \Phi} \Big| S_\lambda - \sum_{i \in H_\lambda} a_i \Big| < \frac{\varepsilon}{2}.$$

The set $H = \bigcup_{\lambda \in \Phi} H_\lambda$ is finite and contains J_0. By the associativity of finite sums, $\sum_{\lambda \in \Phi} \big(\sum_{i \in H_\lambda} a_i \big) = \sum_{i \in H} a_i$. Hence the last displayed inequality can be written as

$$\Big| \sum_{\lambda \in \Phi} S_\lambda - \sum_{i \in H} a_i \Big| < \frac{\varepsilon}{2}.$$

By the definition of S, we have

$$\Big| S - \sum_{i \in H} a_i \Big| < \frac{\varepsilon}{2},$$

because $H \supseteq J_0$. Therefore,

$$\left| S - \sum_{\lambda \in \Phi} S_\lambda \right| = \left| \left(S - \sum_{i \in H} a_i \right) - \left(\sum_{\lambda \in \Phi} S_\lambda - \sum_{i \in H} a_i \right) \right|$$

$$\leq \left| S - \sum_{i \in H} a_i \right| + \left| \sum_{\lambda \in \Phi} S_\lambda - \sum_{i \in H} a_i \right| < \varepsilon,$$

as desired. □

The converse of Theorem 1.16 holds in the following form.

Theorem 1.17. *Let $\{a_i\}_{i \in J}$ be a family of nonnegative numbers and let $J = \bigcup_{\lambda \in \Lambda} J_\lambda$ be a partition of J. If each family $\{a_i\}_{i \in J_\lambda}$ is summable with the sum S_λ, and the family $\{S_\lambda\}_{\lambda \in \Lambda}$ is summable with the sum S, then the family $\{a_i\}_{i \in J}$ is summable and*

$$\sum_{i \in J} a_i = \sum_{\lambda \in \Lambda} \left(\sum_{i \in J_\lambda} a_i \right).$$

Proof. Let K be an arbitrary finite subset of J and let $K_\lambda = K \cap J_\lambda$. It is clear that $K_\lambda \neq \varnothing$ for a finite number of elements $\lambda \in \Lambda$. We have

$$\sum_{i \in K} a_i = \sum_{\substack{\lambda \in \Lambda \\ K_\lambda \neq \varnothing}} \left(\sum_{i \in K_\lambda} a_i \right) \leq \sum_{\substack{\lambda \in \Lambda \\ K_\lambda \neq \varnothing}} S_\lambda \leq S.$$

By Theorem 1.14, the family $\{a_i\}_{i \in J}$ is summable. The desired result follows from Theorem 1.16. □

Let us apply the result of Theorem 1.16 to the case when the index set is a Cartesian product $J = P \times Q$ and the family of real numbers is a summable "double" family $\{a_{pq}\}_{(p,q) \in P \times Q}$. Then, for the partition

$$P \times Q = \bigcup_{p \in P} \{(p, q) : q \in Q\},$$

Theorem 1.16 yields

$$\sum_{p \in P} \left(\sum_{q \in Q} a_{pq} \right) = \sum_{(p,q) \in P \times Q} a_{pq}.$$

Similarly, for the partition $P \times Q = \bigcup_{q \in Q} \{(p, q) : p \in P\}$, we have

$$\sum_{q \in Q} \left(\sum_{p \in P} a_{pq} \right) = \sum_{(p,q) \in P \times Q} a_{pq}.$$

In summary, we obtained the following result.

Theorem 1.18. (Fubini's theorem for sums) *If* $\{a_{pq}\}_{(p,q)\in P\times Q}$ *is a summable double family of nonnegative real numbers, then*

$$\sum_{(p,q)\in P\times Q} a_{pq} = \sum_{p\in P}\left(\sum_{q\in Q} a_{pq}\right) = \sum_{q\in Q}\left(\sum_{p\in P} a_{pq}\right). \qquad (1.4)$$

For a double family, Theorem 1.17 yields another form of Fubini's theorem.

Theorem 1.19. *Let* $\{a_{pq}\}_{(p,q)\in P\times Q}$ *be a double family of nonnegative real numbers such that the family* $\{a_{pq}\}_{q\in Q}$ *is summable for every p in P with the sum* S_p *and the family* $\{S_p\}_{p\in P}$ *is summable. Then the double family* $\{a_{pq}\}_{(p,q)\in P\times Q}$ *is summable. Accordingly, the equalities in (1.4) hold.*

If $P = Q = \mathbb{N}$, we have a convergent double sequence (a_{pq}) which terms form an infinite matrix. This matrix and its row and column sums are shown in the diagram below.

$$
\begin{array}{cccc|c}
a_{11} & a_{12} & \cdots & a_{1q} & \cdots & \sum_q a_{1q} \\
a_{21} & a_{22} & \cdots & a_{2q} & \cdots & \sum_q a_{2q} \\
\vdots & \vdots & & \vdots & & \vdots \\
a_{p1} & a_{p2} & \cdots & a_{pq} & \cdots & \sum_q a_{pq} \\
\vdots & \vdots & & \vdots & & \vdots \\
\hline
\sum_p a_{p1} & \sum_p a_{p2} & \cdots & \sum_p a_{pq} & \cdots & \sum_{p,q} a_{pq}
\end{array}
$$

Notes

The section on sets and functions (Sect. 1.1) serves two purposes. First, it introduces terminology and notation that will be used in the text. Second, it is supplemented by large number of exercises that can be used as warm-up problems at the beginning of the course. The book "Naive Set Theory" [Hal74] by Paul Halmos is an excellent informal introduction to the basic set-theoretic facts.

Sections 1.2 and 1.3 constitute a self-contained (we prove all claims) presentation of very basic facts from real analysis concerning sets and sequences of real numbers and topology of the real line. There are many good books on real analysis; we recommend an elementary introduction [BS11] and a more advanced book [Tao09].

For a general theory of summable families, the reader is referred to [Bou66, III, 5]. The result of Theorem 1.18 holds for absolutely convergent series [Tao09, Sects. 7.4 and 8.2] (see also Exercises 1.53 and 1.54). In the rest of the book, all families of real numbers are at most countable. However, summation over uncountable sets can be found in analysis.

Exercises

1.1. Prove that two sets A and B are equal if and only if

$$A \subseteq B \quad \text{and} \quad B \subseteq A.$$

1.2. Prove identities:

(a) $A \cap B = B \cap A$, $A \cup B = B \cup A$.
(b) $A \cap (B \cap C) = (A \cap B) \cap C$, $A \cup (B \cup C) = (A \cup B) \cup C$.
(c) $A \cap (B \cup C) = (A \cap B) \cup (A \cap C)$, $A \cup (B \cap C) = (A \cup B) \cap (A \cup C)$.
(d) $\complement_X(A \cap B) = (\complement_X A) \cup (\complement_X B)$, $\complement_X(A \cup B) = (\complement_X A) \cap (\complement_X B)$.

1.3. Prove identities:

(a) $(A \setminus B) \cap C = (A \cap C) \setminus (B \cap C)$.
(b) $(A \cup B) \setminus C = (A \setminus C) \cup (B \setminus C)$.
(c) $(A \cap B) \setminus C = (A \setminus C) \cap (B \setminus C)$.
(d) $(A \setminus B) \setminus C = (A \setminus C) \setminus (B \setminus C)$.

1.4. True or false: $(A \setminus B) \setminus C = A \setminus (B \setminus C)$.

1.5. Prove identities:

(a) $A \cap \left(\bigcup_{i \in J} B_i \right) = \bigcup_{i \in J}(A \cap B_i)$, $A \cup \left(\bigcap_{i \in J} B_i \right) = \bigcap_{i \in J}(A \cup B_i)$.

(b) $\complement_X \left[\bigcap_{i \in J} X_i \right] = \bigcup_{i \in J} \left[\complement_X X_i \right]$, $\complement_X \left[\bigcup_{i \in J} X_i \right] = \bigcap_{i \in J} \left[\complement_X X_i \right]$.

(c) $\left(\bigcap_{i \in I} A_i \right) \cup \left(\bigcap_{j \in J} B_j \right) = \bigcap_{(i,j) \in I \times J}(A_i \cup B_j)$.

(d) $\left(\bigcup_{i \in I} A_i \right) \cap \left(\bigcup_{j \in J} B_j \right) = \bigcup_{(i,j) \in I \times J}(A_i \cap B_j)$.

1.6. Let $\{A_{ij}\}_{(i,j) \in \mathbb{N} \times \mathbb{N}}$ be a double indexed family of sets. Prove that

$$\bigcup_{i \in \mathbb{N}} \left(\bigcap_{j \in \mathbb{N}} A_{ij} \right) \subseteq \bigcap_{j \in \mathbb{N}} \left(\bigcup_{i \in \mathbb{N}} A_{ij} \right).$$

Show that the inclusion relation $\subseteq$ cannot be replaced by equality in the above formula. Hint: consider the family of intervals

$$A_{ij} = [j(i-1), ji], \qquad i, j \in \mathbb{N}.$$

1.7. The operation of symmetric difference of two sets A and B is defined by

$$A \triangle B = (A \setminus B) \cup (B \setminus A).$$

Show that

(a) $(A \triangle B) \setminus C = (A \setminus C) \triangle (B \setminus C)$.
(b) $A \triangle B = (A \cup B) \setminus (A \cap B)$.

(c) $(A \triangle B) \triangle C = A \triangle (B \triangle C)$.
(d) $C = A \triangle B$ implies $B = A \triangle C$.

1.8. Let $f : X \to Y$ be a function. Show that

(a) $A \subseteq B \subseteq X$ implies $f(A) \subseteq f(B)$.
(b) $A \subseteq B \subseteq Y$ implies $f^{-1}(A) \subseteq f^{-1}(B)$.

1.9. Let $f : X \to Y$ be a function. Show that the following identities hold for arbitrary sets $A, B \subseteq X$ and $A_1, B_1 \subseteq Y$:

(a) $f(A \cap B) \subseteq f(A) \cap f(B)$.
(b) $f(A \cup B) = f(A) \cup f(B)$.
(c) $f^{-1}(A_1 \cap B_1) = f^{-1}(A_1) \cap f^{-1}(B_1)$.
(d) $f^{-1}(A_1 \cup B_1) = f^{-1}(A_1) \cup f^{-1}(B_1)$.
(e) $f^{-1}(\complement_Y A_1) = \complement_X f^{-1}(A_1)$.
(f) $A \subseteq f^{-1}(f(A))$ and $f(f^{-1}(A_1) \subseteq A_1$.

1.10. Let X be a set and $f : X \to X$ be a function. Let a sequence (A_n) of subsets of X be defined recursively by

$$A_1 = X, \quad A_n = f(A_{n-1}), \quad \text{for } n > 1,$$

and let $A = \bigcap_{n \in \mathbb{N}} A_n$.

(a) Show that $f(A) \subseteq A$.
(b) Show that the inclusion relation $\subseteq$ in part (a) cannot be replaced by equality.

1.11. Let X be a finite set and $f : X \to X$ be a one-to-one function. Prove that f is a bijection.

1.12. Prove that for a function $f : X \to Y$ the following statements are equivalent:

(a) f is one-to-one.
(b) $f^{-1}(f(A)) = A$, for all $A \subseteq X$.
(c) $f(A \cap B) = f(A) \cap f(B)$, for all $A, B \subseteq X$.
(d) $f(A \setminus B) = f(A) \setminus f(B)$, for all $A, B \subseteq X$ such that $B \subseteq A$.

1.13. Let f be a function from $A \times B$ into $B \times A$ defined by

$$f((a, b)) = (b, a).$$

Prove that f is a bijection.

1.14. Let $\{A_i\}_{i \in J}$ be a partition of a set A.

(a) Show that the relation R given by

$$(a, b) \in R \quad \text{if and only if } a, b \in A_j \text{ for some } j \in J,$$

is an equivalence relation on A.

(b) Show that equivalence classes of R from part (a) are exactly elements of the set $\{A_i : i \in J\}$.

1.15. Let a and b be real numbers. Show that

(a) $a \le b$ if and only if $a < b + \varepsilon$ for every $\varepsilon > 0$,
(b) $a \ge b$ if and only if $a > b - \varepsilon$ for every $\varepsilon > 0$.

1.16. Let b be an upper bound of a nonempty set E. Show that

(a) Any $b' > b$ is an upper bound for E.
(b) If c is not an upper bound for E, then there is $x \in E$ such that $x > c$.

1.17. Prove that a nonempty set of real numbers bounded above has a unique least upper bound.

1.18. Let E be a nonempty set bounded below. Show that the set $-E = \{-x : x \in E\}$ is bounded above and

$$- \sup(-E) = \inf E.$$

1.19. Let E be a bounded nonempty set. Show that, for a given number c,

(a) $\sup E \le c$ if and only if $x \le c$ for all $x \in E$.
(b) $\inf E \ge c$ if and only if $x \ge c$ for all $x \in E$.

1.20. Prove that a number u is the upper bound of a nonempty set $E \subseteq \mathbb{R}$ if and only if the condition $t > u$ implies that $t \notin E$.

1.21. Prove that a number u is the supremum of a nonempty set E if and only if u satisfies the two conditions:

 (i) $x \le u$ for all $x \in E$.
(ii) if $v < u$, then there is an $x \in E$ such that $v < x$.

1.22. Let A and B be arbitrary sets of real numbers. Prove that if for any $x \in A$ there is $y \in B$ such that $x \le y$, and for any $y \in B$ there is $x \in A$ such that $y \le x$, then $\sup A = \sup B$.
     Show that the converse is not true.

1.23. Find infimum and supremum of each of the following sets:

(a) $E = \{3, 4, 2, 1, 6, 5, 7, 8\}$.
(b) $E = \{x \in \mathbb{R} : x^3 - x = 0\}$.
(c) $E = [a, b)$, where $a < b$.
(d) $E = \{p/q \in \mathbb{Q} : p^2 < 2q^2,\ p, q > 0\}$.
(e) $E = \{1/n : n \in \mathbb{N}\}$.
(f) $E = \{1 - (-1)^n : n \in \mathbb{N}\}$.
(g) $E = \{1 + (-1)^n/n : n \in \mathbb{N}\}$.

Justify your answers.

1.24. For all the sets in Exercise 1.23 find their maximum and minimum elements or show that these elements do not exist.

1.25. Let E_0 be a nonempty subset of a bounded set E. Show that

$$\inf E \leq \inf E_0 \leq \sup E_0 \leq \sup E.$$

1.26. Let $E \subseteq \mathbb{R}$ be a bounded set and $a \in \mathbb{R}$. We define

$$a + E = \{a + x : x \in E\} \quad \text{and} \quad aE = \{ax : x \in E\}.$$

Show that

(a) $\sup(a + E) = a + \sup E$, $\inf(a + E) = a + \inf E$.
(b) If $a > 0$, then $\sup(aE) = a \sup E$, $\inf(aE) = a \inf E$.
(c) If $a < 0$, then $\sup(aE) = a \inf E$, $\inf(aE) = a \sup E$.

1.27. Prove Theorem 1.5.

1.28. Let A and B be bounded nonempty sets of real numbers. Prove that

$$\sup(A + B) = \sup A + \sup B \quad \text{and} \quad \inf(A + B) = \inf A + \inf B,$$

where $A + B = \{a + b : a \in A, \ b \in B\}$.

1.29. Let (a_n) be a bounded sequence.

(a) Prove that $\liminf a_n = -\limsup(-a_n)$.
(b) Prove that $\liminf a_n \leq \limsup a_n$.
(c) Prove that $\liminf a_n = a$ if and only if for each $\varepsilon > 0$, there are infinitely many indices n for which $a_n < a + \varepsilon$ and only finitely many indices n for which $a_n < a - \varepsilon$.

1.30. Let (a_n) be a convergent sequence. Prove that

$$\lim_{n \to \infty} a_n = \lim_{n \to \infty} \lim_{k \to \infty} \max\{a_n, a_{n+1}, \ldots, a_{n+k}\}.$$

1.31. Prove that for a bounded sequence (a_n),

$$\liminf a_n \leq \limsup a_n.$$

1.32. Prove Theorem 1.5.

1.33. Let (a_n) and (b_n) be two bounded sequences such that $a_n \leq b_n$ for all $n \in \mathbb{N}$. Show that

$$\limsup a_n \leq \limsup b_n \quad \text{and} \quad \liminf a_n \leq \liminf b_n.$$

1.34. Find limit inferior and limit superior of each of the following sequences:

(a) $a_n = 2 - (-1)^n$.
(b) $a_n = \sin(n\pi/2)$.
(c) $a_n = (-1)^n + (-1)^{n+1}/n$.
(d) $a_n = \frac{n^2}{1+n^2} \cos \frac{2n\pi}{3}$.
(e) $a_n = \sqrt[n]{1 + 2^{(-1)^n n}}$.
(f) $u_n = b_n/n$, where (b_n) is a bounded sequence.

Justify your answers.

1.35. Let (a_n) and (b_n) be two bounded sequences. Prove that

$$\liminf a_n + \liminf b_n \leq \liminf(a_n + b_n)$$
$$\leq \limsup(a_n + b_n) \leq \limsup a_n + \limsup b_n.$$

Show that none of the inequality symbols can be replaced by equality.

1.36. For each of the following sets in $\mathbb{R}$, state whether the set is open, closed, or neither.

(a) $\mathbb{Q}$.
(b) $\mathbb{R} \setminus \mathbb{Q}$.
(c) $\bigcup_{k=1}^{\infty} \left(\frac{1}{2k}, \frac{1}{2k-1} \right)$. The set of all rational numbers with denominators that are less than 10^6.
(d) $\bigcup_{k=1}^{\infty} \left[\frac{1}{2k}, \frac{1}{2k-1} \right]$. The set of all rational numbers with denominators that are powers of 2.

1.37. Show that $I_y = I_x$ for any $y \in I_x$ (cf. proof of Theorem 1.7).

1.38. Let $\{(a_i, b_i) : 1 \leq i \leq n\}$ be a finite collection of open intervals that all have a common point $x \in \bigcap_{i=1}^{n}(a_i, b_i)$. Show that the union of these intervals is an open interval containing x.

1.39. Let $\{I_i\}_{i \in J}$ be a family of pairwise disjoint open intervals. Show that these intervals are the component intervals of the union $\bigcup_{i \in J} I_i$.

1.40. Let G be an open subset of the interval $I = (a, b)$, and let (a', b') be a subinterval of (a, b) such that $a < a' < b' < b$. Show that

$$I \setminus [G \cup (a, a') \cup (b', b)] = [a', b'] \cap \complement G.$$

1.41. Show that every closed set is the intersection of a countable family of open sets.

1.42. Show that an open interval (a, b) cannot be represented as the union of a countable family of mutually disjoint closed sets. [Knu76].

1.43. Show that the set of irrational points in $[0,1]$ cannot be represented as the union of a countable family of closed sets.
(Hint: use Baire's Category Theorem [Kre78, p. 247].)

1.44. Prove formula (1.3) (cf. Exercise 1.26).

1.45. Show that the Cantor set contains no open interval.

1.46. Show that the number $\frac{1}{4}$ is in the Cantor set.

1.47. Let $\{a_i\}_{i \in J}$ be an unbounded family of nonnegative numbers. Show that this family is not summable. Give an example of a bounded family of nonnegative numbers which is not summable.

1.48. Let S be the sum of a summable family $\{a_i\}_{i \in J}$. Show that for any given $\varepsilon > 0$ and a finite subset A of J, there is a finite subset $A(\varepsilon)$ of J containing A such that $|S - S_{A(\varepsilon)}| < \varepsilon$.

1.49. Let $\{a_i\}_{i \in J}$ be a countable family of nonnegative real numbers and let $f : \mathbb{N} \to J$ be a bijection. Prove that the series $\sum_{i=1}^{\infty} a_{f(i)}$ converges to S if and only if the family $\{a_i\}_{i \in J}$ is summable to S.

1.50. Show that the series $\sum \frac{(-1)^n}{n}$ is convergent, but the sequence $\left(\frac{(-1)^n}{n}\right)$ is not summable.

1.51. Let $\sum_{i,j \in \mathbb{N}} a_{ij}$ be a convergent double series with nonnegative terms. Prove that the series $\sum_{i,j \in \mathbb{N}} a_{ji}$ converges to the same sum $S = \sum_{i,j \in \mathbb{N}} a_{ij}$. (Hint: use the result of Exercise 1.13.)

1.52. Prove Theorem 1.15.

1.53. Use the power series expansion of $\ln(1 + x)$ to show that

$$1 - \frac{1}{2} + \frac{1}{3} - \frac{1}{4} + \cdots + \frac{1}{2k-1} - \frac{1}{2k} + \cdots = \ln 2.$$

Let us rearrange terms of this conditionally convergent series, so two negative terms follow each positive term:

$$1 - \frac{1}{2} - \frac{1}{4} + \frac{1}{3} - \frac{1}{6} - \frac{1}{8} + \cdots + \frac{1}{2k-1} - \frac{1}{4k-2} - \frac{1}{4k} + \cdots .$$

Show that the sum of the new series is one-half of the sum of the original series. (Hint: $\frac{1}{2k-1} - \frac{1}{4k-2} - \frac{1}{4k} = \frac{1}{2}\left(\frac{1}{2k-1} - \frac{1}{2k}\right)$.)

1.54. Let (a_{ij}) for $i, j \in \mathbb{N}$ be a double sequence given by the matrix

$$\begin{pmatrix} 1 & 0 & 0 & 0 & \cdots \\ -1 & 1 & 0 & 0 & \cdots \\ 0 & -1 & 1 & 0 & \cdots \\ 0 & 0 & -1 & 1 & \cdots \\ 0 & 0 & 0 & -1 & \cdots \\ \vdots & \vdots & \vdots & \vdots & \ddots \end{pmatrix}.$$

Show that

$$\sum_{i=1}^{\infty} \left(\sum_{j=1}^{\infty} a_{ij} \right) \neq \sum_{j=1}^{\infty} \left(\sum_{i=1}^{\infty} a_{ij} \right).$$

1.55. Compute the double sum

$$\sum_{\substack{k>1 \\ n>1}} \frac{1}{n^k}.$$

1.56. Let a and b be two numbers such that $0 \leq a < 1$, $0 \leq b < 1$. Show that the double sequence $\{a^m b^n\}_{(m,n) \in \mathbb{N} \times \mathbb{N}}$ is summable.

1.57. Show that for each $p > 1$, the sequence (n^{-p}) is summable. (Hint: Show that $S_{2^{n+1}} - S_{2^n} < 2^n (2^n)^{-p}$, and therefore, by adding these inequalities,

$$S_{2^n} < \frac{1}{1 - 2^{1-p}}.)$$

2

Lebesgue Measure

In this chapter we present elements of Lebesgue's theory of measure and measurable functions.

The notion of measure of a set of real numbers generalizes the concept of length of an interval. Because open sets have a very simple structure—they are at most countable unions of open intervals—we begin by defining the measure of a bounded open set in Sect. 2.1. Closed sets are complements of open sets, therefore it is possible to extend the concept of measure to bounded closed sets as it is presented in Sect. 2.2. For an arbitrary bounded set of real numbers, we define its outer and inner measures by "approximating" the set by, respectively, open sets containing the set and closed sets that are contained in the set. These measures are investigated in Sect. 2.3. The set is called measurable if its outer and inner measures are equal. The common value of these two measures is the Lebesgue measure of a measurable set. Main properties of measurable sets and measure are introduced in Sect. 2.4. In particular, it is shown there that the set of measurable sets is closed under countable unions and intersections and that measure is a countably additive function on the set of measurable sets. Another important property of measure, its translation invariance, is established in Sect. 2.5. An example of a nonmeasurable set is given in Sect. 2.6, where some general properties of the class of measurable sets are discussed.

The concept of a measurable function plays a vital role in real analysis. These functions are introduced and their basic properties are investigated in Sect. 2.7. Various types of convergence of sequences of measurable functions are discussed in Sect. 2.8.

Occasionally, we consider unbounded families of nonnegative numbers. If $\{a_i\}_{i \in J}$ is such a family, then we set $\sum_{i \in J} a_i = \infty$ (cf. Exercise 1.47) and assume that $a < \infty$ for all $a \in \mathbb{R}$ (see conventions in Sect. 1.2). Here and in the rest of the book, we write ∞ for $+\infty$.

S. Ovchinnikov, *Measure, Integral, Derivative: A Course on Lebesgue's Theory*,   27
Universitext, DOI 10.1007/978-1-4614-7196-7_2,
© Springer Science+Business Media New York 2013

2.1 The Measure of a Bounded Open Set

Definition 2.1. *The measure of an open interval $I = (a, b)$ is its length,*

$$m(I) = b - a.$$

Recall (cf. Theorem 1.7) that an open set is the union of a family of pairwise disjoint open intervals. The following theorem is instrumental.

Theorem 2.1. *Let $\{I_i\}_{i \in J}$ be a family of pairwise disjoint open intervals that are contained in an open interval $I = (a, b)$. If J is a finite or countable set, then the family $\{m(I_i)\}_{i \in J}$ is summable and $\sum_{i \in J} m(I_i) \leq m(I)$.*

Note that the set of indexes J is at most countable, because the intervals I_i are pairwise disjoint.

Proof. First, suppose that J is a finite set and let $n = |J|$. We number the intervals I_i from left to right, so $I_i = (a_i, b_i)$, $1 \leq i \leq n$, and

$$a \leq a_1 < b_1 \leq a_2 < b_2 \leq \cdots \leq a_n < b_n \leq b$$

(cf. Fig. 2.1).

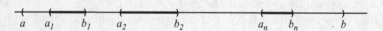

Figure 2.1. Proof of Theorem 2.1

Clearly,

$$(b_1 - a_1) + (b_2 - a_2) + \cdots + (b_n - a_n)$$
$$\leq (a_1 - a) + (b_1 - a_1) + (a_2 - b_1) + \cdots + (b_n - a_n) + (b - b_n)$$
$$= b - a.$$

Hence, $\sum_{i \in J} m(I_i) \leq m(I)$.

The claim follows immediately from Theorem 1.14, if J is a countable set. $\square$

The result of Theorem 2.1 justifies the following definition.

Definition 2.2. *Let G be a nonempty bounded open set and let $\{I_i\}_{i \in J}$ be the family of its component intervals. The measure $m(G)$ of the set G is the sum of measures of its component intervals:*

$$m(G) = \sum_{i \in J} m(I_i). \tag{2.1}$$

We also set $m(\varnothing) = 0$.

Thus, m is a real-valued function on the family of all bounded open subsets of $\mathbb{R}$. In the rest of the section we establish some useful properties of this function.

Theorem 2.2. *If an open set G_1 is a subset of a bounded open set G_2, then $m(G_1) \leq m(G_2)$. In words, measure is a monotone function on the set of bounded open subsets of real numbers.*

Proof. Let $\{I_i\}_{i \in J}$ be the family of component intervals of the set G_1. By Corollary 1.1, each I_i is contained in a unique component interval of the set G_2. Let us partition the set J into a family of subsets $\{J_\lambda\}_{\lambda \in \Lambda}$ in such a way that indices i and j are in the same set J_λ if and only if the intervals I_i and I_j are subsets of the same component interval I'_λ of G_2. By Theorem 1.16 (the associativity property of addition),

$$m(G_1) = \sum_{i \in J} m(I_i) = \sum_{\lambda \in \Lambda} \Big(\sum_{i \in J_\lambda} m(I_i) \Big),$$

and, by Theorem 2.1,

$$\sum_{i \in J_\lambda} m(I_i) \leq m(I'_\lambda).$$

Hence,

$$m(G_1) = \sum_{\lambda \in \Lambda} \Big(\sum_{i \in J_\lambda} m(I_i) \Big) \leq \sum_{\lambda \in \Lambda} m(I'_\lambda).$$

By Corollary 1.2, we have

$$\sum_{\lambda \in \Lambda} m(I'_\lambda) \leq m(G_2),$$

inasmuch as the family $\{I'_\lambda\}_{\lambda \in \Lambda}$ is a subfamily of the family of component intervals of G_2. Thus, $m(G_1) \leq m(G_2)$.   □

The next theorem asserts that measure is a countably additive function on bounded open sets (cf. (2.1)).

Theorem 2.3. *Let a bounded open set G be the union of a finite or countable family of pairwise disjoint open sets $\{G_i\}_{i \in J}$. Then*

$$m(G) = \sum_{i \in J} m(G_i).$$

Proof. For a given $i \in J$, let $G_i = \bigcup_{j \subset J_i} I_j^{(i)}$, where $\{I_j^{(i)}\}_{j \in J_i}$ is the family of component intervals of the set G_i. By Theorem 1.8, the intervals $I_j^{(i)}$ for $j \in J_i$, $i \in J$, are precisely the component intervals of the set G. By the associativity property of summable families, we have

$$m(G) = \sum_{\substack{j \in J_i \\ i \in J}} m(I_j^{(i)}) = \sum_{i \in J} \left(\sum_{j \in J_i} m(I_j^{(i)}) \right) = \sum_{i \in J} m(G_i)$$

(cf. Theorem 1.16). □

If we drop the assumption that the sets in the family $\{G_i\}_{i \in J}$ in Theorem 2.3 are pairwise disjoint, then we have a weaker property of measure:

$$m(G) \le \sum_{i \in J} m(G_i). \tag{2.2}$$

(Recall that we set $\sum_{i \in J} m(G_i) = \infty$ if the family $\{m(G_i)\}_{i \in J}$ is unbounded.) In order to prove this claim, we need two lemmas.

Lemma 2.1. *If a closed bounded interval $[a, b]$ is contained in the finite union of open intervals, then the length of $[a, b]$ is less than the sum of lengths of the open intervals.*

Proof. The proof is by induction on the number n of open intervals. The claim is trivial if $n = 1$. Let us assume that it holds for some n, and let $\{(a_i, b_i)\}_{1 \le i \le n+1}$ be a family of $n+1$ open intervals such that their union contains $[a, b]$. Without loss of generality, we may assume that $b \in (a_{n+1}, b_{n+1})$. If $a_{n+1} < a$, we are done. Otherwise,

$$[a, a_{n+1}] \subseteq \bigcup_{i=1}^{n} (a_i, b_i).$$

By the induction hypothesis,

$$a_{n+1} - a < \sum_{i=1}^{n} (b_i - a_i).$$

Therefore,

$$b - a = (b - a_{n+1}) + (a_{n+1} - a)$$
$$< (b_{n+1} - a_{n+1}) + \sum_{i=1}^{n} (b_i - a_i) = \sum_{i=1}^{n+1} (b_i - a_i),$$

and the result follows. □

Lemma 2.2. *If an open interval $I = (a, b)$ is the union of a finite or countable family of open sets $\{G_i\}_{i \in J}$, then*

$$m(I) \le \sum_{i \in J} m(G_i).$$

Proof. The result is trivial if $\sum_{i \in J} m(G_i) = \infty$, so we assume that the family $\{m(G_i\}_{i \in J}$ is summable.

For a given $i \in J$, let $\{I_k^{(i)}\}_{k \in J_i}$ be the family of component intervals of the set G_i. Let us select $0 < \varepsilon < (b-a)$ and consider the closed interval $[a + \varepsilon/2, b - \varepsilon/2]$. The family of open intervals

$$\{I_k^{(i)} : k \in J_i, i \in J\}$$

covers this interval. By Heine–Borel theorem (Theorem 1.2), this family contains a finite subfamily

$$\{I_{k_s}^{(i_s)} : k_s \in J_{i_s}, i_s \in J, s = 1, \ldots, n\}$$

covering the same interval (cf. Fig. 2.2). By Lemma 2.1,

$$b - a - \varepsilon < \sum_{s=1}^{n} m(I_{k_s}^{(i_s)}).$$

It is clear that the families $\{m(I_k^{(i)})\}_{k \in J_i}$, $i \in J$ are summable. Therefore,

$$b - a - \varepsilon < \sum_{s=1}^{n} m(I_{k_s}^{(i_s)}) = \sum_{i_s \in J} \left(\sum_{k_s \in J_{i_s}} m(I_{k_s}^{(i_s)}) \right)$$

$$\leq \sum_{i \in J} \left(\sum_{k \in J_i} m(I_k^{(i)}) \right) = \sum_{i \in J} m(G_i).$$

Because this inequality holds for an arbitrary small number $\varepsilon > 0$, we obtained the desired result. □

The drawing in Fig. 2.2 illustrates the proof of Lemma 2.2.

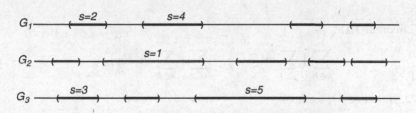

Figure 2.2. Proof of Lemma 2.2

In this example, $J = \{1, 2, 3\}$ and

$$J_1 = \{1, 2, 3, 4\}, \quad J_2 = \{1, 2, 3, 4, 5\}, \quad J_3 = \{1, 2, 3, 4\}.$$

There are $n = 5$ selected intervals labeled by the values of the variable s in Fig. 2.2. It can be easily seen that

$$(i_s) = (2, 1, 3, 1, 3) \quad \text{and} \quad (k_s) = (2, 1, 1, 2, 3).$$

Now we prove (2.2).

Theorem 2.4. *If a bounded open set G is the union of a finite or countable family of open sets $\{G_i\}_{i \in J}$, then*

$$m(G) \leq \sum_{i \in J} m(G_i).$$

Proof. The result is trivial if $\sum_{i \in J} m(G_i) = \infty$, so we assume that the family $\{m(G_i)_{i \in J}$ is summable.

We denote by $\{I_k\}_{k \in K}$ the family of component intervals of the set G and consider open sets

$$E_{ik} = G_i \cap I_k, \qquad i \in J, \ k \in K.$$

By Theorem 2.2, we have $m(E_{ik}) \leq m(G_i)$ for any given $i \in J, k \in K$. Because the family $\{m(G_i)\}_{i \in J}$ is summable, the family $\{m(E_{ik})\}_{i \in J}$ is summable for any $k \in K$. By Lemma 2.2, we have

$$m(I_k) \leq \sum_{i \in J} m(E_{ik}), \qquad \text{for every } k \in K. \tag{2.3}$$

Furthermore, for $i \in J$,

$$G_i = G_i \cap \left(\bigcup_{k \in K} I_k \right) = \bigcup_{k \in K} (G_i \cap I_k) = \bigcup_{k \in K} E_{ik},$$

where the open sets in the right union are pairwise disjoint. By Theorem 2.3,

$$m(G_i) = \sum_{k \in K} m(E_{ik}), \qquad \text{for every } i \in J. \tag{2.4}$$

By Theorem 1.19,

$$\sum_{k \in K} \left(\sum_{i \in J} m(E_{ik}) \right) = \sum_{i \in J} \left(\sum_{k \in K} m(E_{ik}) \right).$$

By (2.3) and (2.4), we have

$$m(G) = \sum_{k \in K} m(I_k) \leq \sum_{k \in K} \left(\sum_{i \in J} m(E_{ik}) \right)$$
$$= \sum_{i \in J} \left(\sum_{k \in K} m(E_{ik}) \right) = \sum_{i \in J} m(G_i),$$

which yields the desired result. $\qquad \square$

2.2 The Measure of a Bounded Closed Set

The result of Theorem 1.10 justifies the following definition.

Definition 2.3. *Let F be a nonempty bounded closed set and let $a = \inf F$, $b = \sup F$. The measure of F is given by*

$$m(F) = (b - a) - m([a, b] \setminus F).$$

By Theorem 1.11, the class of nonempty proper closed subsets of $\mathbb{R}$ is disjoint from the class of open subsets. Definition 2.3 extends the function m from the class of bounded open sets to the class of bounded nonempty closed sets. We denote this extended function by the same symbol m.

By Theorem 2.1, the measure of the open set $[a, b] \setminus F$ is less than the length of the interval $[a, b]$. Hence, the function m is nonnegative on bounded open and closed sets.

Example 2.1. Let $F = [a, b]$, so $[a, b] \setminus F = \varnothing$. Therefore,

$$m([a, b]) = b - a.$$

In particular, the measure of a singleton is zero.

The following lemma is instrumental. The result of this lemma will be used implicitly in the rest of this section.

Lemma 2.3. *Let $I = (A, B)$ be an open interval containing a closed set F. Then*

$$m(F) = m(I) - m(I \setminus F).$$

Note that $I \setminus F = I \cap \complement F$ is a bounded open set. Thus $m(I \setminus F)$ is well defined.

Proof. As before, let $a = \inf F$, $b = \sup F$. The sets $[a, b] \setminus F$ and $I \setminus [a, b]$ are open and disjoint (cf. Fig. 2.3).

Figure 2.3. Proof of Lemma 2.3

Furthermore (cf. Exercise 2.3),

$$I \setminus F = ([a, b] \setminus F) \cup (I \setminus [a, b]).$$

By Theorem 2.3 and Definition 2.3,

$$m(I \setminus F) = m([a,b] \setminus F) + m(I \setminus [a,b])$$
$$= [(b-a) - m(F)] + [(B-A) - (b-a)]$$
$$= (B-A) - m(F),$$

and the desired result follows. □

Example 2.2. Let $F = \bigcup_{i \in J} [a_i, b_i]$ be the union of a finite family of pairwise disjoint closed intervals (see Fig. 2.4). We may assume that the intervals are

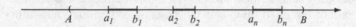

Figure 2.4. Example 2.2

numerated in such a way that $a_1 \le b_1 < a_2 \le b_2 < \cdots < a_n \le b_n$, where $n = |J|$ is the cardinality of J. Let $I = (A, B)$ be an open interval containing F. It is clear that

$$I \setminus F = (A, a_1) \cup (b_1, a_2) \cup \cdots \cup (b_k, a_{k+1}) \cup \cdots \cup (b_n, B).$$

Hence, by Theorem 2.3,

$$m(I \setminus F) = (a_1 - A) + (a_2 - b_1) + \cdots + (a_n - b_{n-1}) + (B - b_n)$$
$$= (B - A) - (b_1 - a_1) - \cdots - (b_n - a_n)$$
$$= m(I) - \sum_{i=1}^{n} m([a_i, b_i]).$$

Hence, $m(F) = \sum_{i=1}^{n} m([a_i, b_i])$.

By Theorem 2.2, the function m is monotone on the set of bounded open sets of real numbers. The following three lemmas demonstrate that the extended function m is also monotone.

Lemma 2.4. *Let a closed set F_1 be a subset of a bounded closed set F_2. Then*

$$m(F_1) \le m(F_2).$$

Proof. Let $I = (A, B)$ be an open interval such that $F_1 \subseteq F_2 \subseteq I$. The sets $I \setminus F_i = I \cap \complement F_i$, $i \in \{1, 2\}$, are open and bounded. Clearly, $I \setminus F_2 \subseteq I \setminus F_1$. By Theorem 2.2,

$$m(I \setminus F_2) \le m(I \setminus F_1),$$

that is,

$$m(I) - m(F_2) \le m(I) - m(F_1).$$

Therefore, $m(F_1) \le m(F_2)$. □

Lemma 2.5. *Let F be a closed subset of a bounded open set G. Then*

$$m(F) \leq m(G).$$

Proof. Let $I = (A, B)$ be an open interval containing G. Then

$$I = G \cup (I \setminus F)$$

(cf. Exercise 2.6), where the sets G and $I \setminus F = I \cap CF$ are open. By Theorem 2.4,

$$m(I) \leq m(G) + m(I \setminus F),$$

that is,

$$B - A \leq m(G) + (B - A) - m(F).$$

Hence, $m(F) \leq m(G)$. □

Lemma 2.6. *Let G be an open subset of a bounded closed set F. Then*

$$m(G) \leq m(F).$$

Proof. Let $a = \inf F$ and $b = \sup F$. The sets G and $[a, b] \setminus F$ are open and disjoint. It is clear that

$$G \cup ([a, b] \setminus F) \subseteq (a, b).$$

Therefore, by Theorems 2.3 and 2.2,

$$m(G) + m([a, b] \setminus F) \leq b - a,$$

that is, by the definition of $m(F)$,

$$m(G) + (b - a) - m(F) \leq b - a,$$

and the result follows. □

In summary, we have the following theorem.

Theorem 2.5. *Let U and V be bounded sets of real numbers such that each of these two sets is either open or closed. Then*

$$U \subseteq V \qquad implies \qquad m(U) \leq m(V).$$

It follows immediately from Theorem 2.2 that the measure of a bounded open set G is the greatest lower bound of measures of bounded open sets containing G:

$$m(G) = \inf\{m(G') : G \subseteq G', \ G' \text{ is open and bounded}\}.$$

Similar results are established in the next three theorems.

Theorem 2.6. *Let G be a bounded open set. Then*

$$m(G) = \sup\{m(F) : F \subseteq G, \ F \text{ is closed}\}.$$

Proof. By Theorem 2.5, $m(F) \leq m(G)$ for any closed subset F of the set G. According to the approximation property of supremum (cf. Sect. 1.2), we need to show that for a given $\varepsilon > 0$ there is a closed set $F \subseteq G$ such that $m(F) > m(G) - \varepsilon$.

The set G is the union of at most countable family of its component intervals, $G = \bigcup_{i \in J}(a_i, b_i)$, and the measure of G is the sum of lengths of these intervals,

$$m(G) = \sum_{i \in J}(b_i - a_i).$$

Therefore there is a finite set $J_0 \subseteq J$ such that

$$\sum_{i \in J_0}(b_i - a_i) > m(G) - \varepsilon/2.$$

Let $n = |J_0|$ be the cardinality of the set J_0. For every $i \in J_0$, we choose an interval $[a_i', b_i']$ so that

$$[a_i', b_i'] \subseteq (a_i, b_i) \quad \text{and} \quad (b_i' - a_i') > (b_i - a_i) - \frac{\varepsilon}{2n}$$

(cf. Exercise 2.7) and define a closed set F by

$$F = \bigcup_{i \in J_0} [a_i', b_i'].$$

We have (cf. Example 2.2),

$$m(F) = \sum_{i \in J_0}(b_i' - a_i') > \left[\sum_{i \in J_0}(b_i - a_i)\right] - \varepsilon/2 > m(G) - \varepsilon,$$

which is the desired result. □

Theorem 2.7. *Let F be a bounded closed set. Then*

$$m(F) = \inf\{m(G) : F \subseteq G, \ G \text{ is open and bounded}\}.$$

Proof. Let $I = (A, B)$ be an open interval containing the set F. The set $I \setminus F$ is open and bounded. By Theorem 2.6, for a given $\varepsilon > 0$, there is a closed set $F_0 \subseteq I \setminus F$ such that

$$m(F_0) > m(I \setminus F) - \varepsilon = (B - A) - m(F) - \varepsilon.$$

The set $G_0 = I \setminus F_0$ is open and contains the set F. Therefore, by the above inequality,

$$m(G_0) = m(I \setminus F_0) - (B - A) - m(F_0)$$
$$< (B - A) - (B - A) + m(F) + \varepsilon = m(F) + \varepsilon.$$

By Theorem 2.5, $m(G_0) \geq m(F)$. Now the result follows from the approximation property of infimum. □

From Lemma 2.4 we immediately obtain the following result.

Theorem 2.8. *Let F be a bounded closed set. Then*

$$m(F) = \sup\{m(F') : F' \subseteq F, \ F' \ is \ closed\}.$$

We summarize these results in the following theorem which justifies definitions of the inner and outer measures in Sect. 2.3.

Theorem 2.9. *Let U be a bounded open or closed set. Then*

$$m(U) = \inf\{m(G) : U \subseteq G, \ G \ is \ open \ and \ bounded\}$$
$$= \sup\{m(F) : F \subseteq U, \ F \ is \ closed\}.$$

By Theorem 2.3, the function m is countably additive on the class of bounded open sets. For closed sets, we need only the finite additivity property in Sect. 2.3.

Theorem 2.10. *Let $\{F_1, \ldots, F_n\}$ be a finite collection of pairwise disjoint bounded closed sets and let $F = \bigcup_{i=1}^n F_i$. Then*

$$m(F) = \sum_{i=1}^n m(F_i).$$

Proof. The proof is by induction. For $n = 2$ we have two bounded closed sets F_1 and F_2 with $F_1 \cap F_2 = \varnothing$.

By Theorem 2.7, for a given $\varepsilon > 0$, there is an open set G containing F such that

$$m(F) > m(G) - \varepsilon.$$

By Theorem 1.12, there are open sets $G_1 \supseteq F_1$ and $G_2 \supseteq F_2$ such that $G_1 \cap G_2 = \varnothing$. Clearly, $F_i \subseteq G \cap G_i$ for $i \in \{1, 2\}$ and

$$(G_1 \cap G) \cap (G_2 \cap G) = \varnothing, \quad (G_1 \cap G) \cup (G_2 \cap G) \subseteq G.$$

Therefore, by Theorem 2.5,

$$m(F_1) + m(F_2) \leq m(G \cap G_1) + m(G \cap G_2)$$
$$= m((G_1 \cap G) \cup (G_2 \cap G)) \leq m(G) < m(F) + \varepsilon.$$

Because ε is an arbitrary positive number, we have

$$m(F_1) + m(F_2) \leq m(F).$$

To prove the opposite inequality, we consider open sets $G_i \supseteq F_i$ such that $m(F_i) > m(G_i) - \frac{\varepsilon}{2}$, for $i \in \{1, 2\}$. These sets exist by Theorem 2.7. Then we have

$$m(F_1) + m(F_2) > m(G_1) + m(G_2) - \varepsilon.$$

From this inequality we obtain, by applying Theorems 2.5 and 2.4,

$$m(F_1 \cup F_2) \leq m(G_1 \cup G_2) \leq m(G_1) + m(G_2)$$
$$< m(F_1) + m(F_2) + \varepsilon.$$

Hence, $m(F) \leq m(F_1) + m(F_2)$, and we obtain the desired result:

$$m(F) = m(F_1) + m(F_2).$$

For the induction step, let $\{F_1, \ldots, F_n\}$ be a finite collection of mutually disjoint bounded closed sets and let $F = \bigcup_{i=1}^{n} F_i$. By the induction hypothesis, for $F' = \bigcup_{i=1}^{n-1} F_i$, we have

$$m(F') = \sum_{i=1}^{n-1} m(F_i).$$

Clearly, $F = F' \cup F_n$ and $F' \cap F_n = \varnothing$. As we proved before,

$$m(F) = m(F') + m(F_n).$$

This implies

$$m(F) = \sum_{i=1}^{n} m(F_i),$$

which completes the proof.   □

2.3 Inner and Outer Measures

Definition 2.4. *Let E be a bounded set.*

(i) *The* outer measure $m^*(E)$ *of E is the greatest lower bound of measures of all bounded open sets G containing the set E:*

$$m^*(E) = \inf\{m(G) : G \supseteq E, \ G \text{ is open and bounded}\}.$$

(ii) *The* inner measure $m_*(E)$ *of E is the least upper bound of measures of closed sets F contained in the set E:*

$$m_*(E) = \sup\{m(F) : F \subseteq E, \ F \text{ is closed}\}.$$

It is clear that

$$0 \leq m^*(E) < \infty \quad \text{and} \quad 0 \leq m_*(E) < \infty$$

for any bounded set E. By Theorem 2.9,

$$m^*(G) = m_*(G) = m(G) \quad \text{and} \quad m^*(F) = m_*(F) = m(F),$$

for any bounded open set G and any bounded closed set F.

Theorem 2.11. *For every bounded set E,*

$$m_*(E) \leq m^*(E).$$

Proof. Let G be a bounded open set containing E. For any closed subset F of the set E, we have $F \subseteq E \subseteq G$. Therefore, by Theorem 2.5, $m(F) \leq m(G)$, that is, $m(G)$ is an upper bound of the family $\{m(F)\}_{F \subseteq G}$. Hence,

$$m_*(E) = \sup_{F \subseteq E} \{m(F)\} \leq \sup_{F \subseteq G} \{m(F)\} = m(G).$$

It follows that $m_*(E)$ is a lower bound for the measures of bounded open sets containing E. Thus,

$$m_*(E) \leq \inf_{G \supseteq E} \{m(G)\} = m^*(E),$$

which proves the claim. $\qquad \square$

The monotonicity property of inner and outer measures is established in the next theorem.

Theorem 2.12. *Let U and V be bounded sets. If $U \subseteq V$, then*

$$m_*(U) \leq m_*(V) \quad \text{and} \quad m^*(U) \leq m^*(V).$$

Proof. We prove the first inequality leaving the second one as an exercise (cf. Exercise 2.8).

Let us consider sets of real numbers:

$$A = \{m(F) : F \text{ is a closed subset of } U\},$$
$$B = \{m(F) : F \text{ is a closed subset of } V\}.$$

Clearly, $A \subseteq B$. Therefore, $m_*(U) = \sup A \leq \sup B = m_*(V)$ (cf. Exercise 1.25). $\qquad \square$

We obtain a generalization of Theorem 2.4 as the result of the following theorem (countable subadditivity of the outer measure).

Theorem 2.13. *If a bounded set E is the union of a finite or countable family of sets $\{E_i\}_{i\in J}$,*

$$E = \bigcup_{i\in J} E_i,$$

then

$$m^*(E) \le \sum_{i\in J} m^*(E_i).$$

Proof. First, we consider the case when J is a countable set. The result is trivial if $\sum_{i\in J} m^*(E_i) = \infty$, so we assume that the family $\{m^*(E_i)\}_{i\in J}$ is summable.

Let $\varepsilon > 0$ be a given number and $f : J \to \mathbb{N}$ be a bijection. By the definition of the outer measure, for each $i \in J$ there is an open set $G_i \supseteq E_i$ such that

$$m(G_i) < m^*(E_i) + \frac{\varepsilon}{2^{f(i)}}.$$

Let (a,b) be an open interval containing the set E. The set

$$(a,b) \cap \bigcup_{i\in J} G_i = \bigcup_{i\in J} [(a,b) \cap G_i]$$

is open and contains E. Therefore, by Theorems 2.12, 2.4, and 2.2,

$$m^*(E) \le m\Big(\bigcup_{i\in J} [(a,b) \cap G_i] \Big) \le \sum_{i\in J} m[(a,b) \cap G_i]$$

$$\le \sum_{i\in J} m(G_i) \le \sum_{i\in J} \Big(m^*(E_i) + \frac{\varepsilon}{2^{f(i)}} \Big) = \sum_{i\in J} m^*(E_i) + \varepsilon.$$

(cf. Exercise 1.49). Because ε is an arbitrary positive number, we have

$$m^*(E) \le \sum_{i\in J} m^*(E_i).$$

The case when J is a finite set is reduced to the previous one by adding countably many empty sets to the family $\{E_i\}_{i\in J}$. $\qquad\square$

For the inner measure, we have a weaker result.

Theorem 2.14. *If a bounded set E is the union of a finite or countable family of pairwise disjoint sets $\{E_i\}_{i\in J}$,*

$$E = \bigcup_{i\in J} E_i, \qquad E_i \cap E_j = \varnothing \text{ for } i \ne j,$$

then

$$m_*(E) \ge \sum_{i\in J} m_*(E_i).$$

Proof. As in the proof of the previous theorem, it suffices to consider the case when J is a countable set.

Let J_0 be a finite subset of J of cardinality n, and let $\varepsilon > 0$ be a given number. By the definition of the inner measure, for each $i \in J_0$, there is a closed set $F_i \subseteq E_i$ such that

$$m(F_i) > m_*(E_i) - \frac{\varepsilon}{n}.$$

It is clear that the sets F_i are pairwise disjoint and their union $\bigcup_{i \in J_0} F_i$ is a closed subset of the set E. By the definition of the inner measure and Theorems 2.12 and 2.10,

$$m_*(E) \geq m\Big(\bigcup_{i \in J_0} F_i \Big) = \sum_{i \in J_0} m(F_i)$$
$$> \sum_{i \in J_0} \Big(m_*(E_i) - \frac{\varepsilon}{n} \Big) = \sum_{i \in J_0} m_*(E_i) - \varepsilon.$$

Because ε is an arbitrary positive number, it follows that

$$\sum_{i \in J_0} m_*(E_i) \leq m_*(E).$$

By Theorem 1.14, the family $\{m_*(E_i)\}_{i \in J}$ is summable and

$$\sum_{i \in J} m_*(E_i) \leq m_*(E),$$

which proves the assertion of the theorem. $\square$

Note that, unlike in the case of the outer measure, the claim of the previous theorem fails to hold if the sets E_i's are not pairwise disjoint. Indeed, let $E_1 = [0, 1]$ and $E_2 = [0, 2]$. Then

$$m_*(E_1 \cup E_2) = 2 < 3 = m_*(E_1) + m_*(E_2).$$

The last theorem of this section establishes an important property of inner and outer measures which is used several times in the rest of this chapter.

Theorem 2.15. *Let E be a bounded set and $I = (a, b)$ be an open interval containing E. Then*
$$m^*(E) + m_*(I \setminus E) = m(I).$$

Proof. By the definition of the outer measure, for a given $\varepsilon > 0$, there is an open set $G_0 \supseteq E$ such that $m(G_0) < m^*(E) + \varepsilon$. Let (a', b') be a subinterval of (a, b) such that $0 < a' - a < \varepsilon$, $0 < b - b' < \varepsilon$, and let

$$G = (I \cap G_0) \cup (a, a') \cup (b', b).$$

By Theorems 2.4 and 2.2,

$$m(G) \leq m(I \cap G_0) + (a' - a) + (b - b')$$
$$\leq m(G_0) + (a' - a) + (b - b') < m^*(E) + 3\varepsilon.$$

The set $F = I \setminus G$ is closed, because

$$F = [a', b'] \cap \complement G$$

(cf. Exercise 1.40). Since $E \subseteq G$, we have $F = I \setminus G \subseteq I \setminus E$. Hence,

$$m_*(I \setminus E) \geq m(F) = m(I) - m(G) > m(I) - m^*(E) - 3\varepsilon.$$

Because ε is an arbitrary positive number, we have the inequality

$$m^*(E) + m_*(I \setminus E) \geq m(I).$$

In order to obtain the reverse inequality,

$$m^*(E) + m_*(I \setminus E) \leq m(I),$$

let us select a closed set F such that $F \subseteq I \setminus E$ and

$$m(F) > m_*(I \setminus E) - \varepsilon,$$

where $\varepsilon > 0$ is a given number. The set $G = I \setminus F$ is a bounded open set containing the set E. Therefore,

$$m^*(E) \leq m(G) = m(I) - m(F) < m(I) - m_*(I \setminus E) + \varepsilon,$$

which yields the desired inequality, because ε is an arbitrary positive number. $\square$

2.4 Measurable Sets

Definition 2.5. *A bounded set E is said to be* measurable *if its outer and inner measures are equal. The* measure $m(E)$ *of a measurable set E is the common value of its outer and inner measures:*

$$m(E) = m^*(E) = m_*(E).$$

The following theorem justifies the notation m in the above definition.

Theorem 2.16. (i) *A bounded open set is measurable and its measure is the same as defined in Sect. 2.1.*

(ii) *A bounded closed set is measurable and its measure is the same as defined in Sect. 2.2.*

Proof. The claims follow immediately from Theorem 2.9 (cf. the second paragraph in Sect. 2.3). □

Theorem 2.17. *Let E be a subset of an open bounded interval I. Then E is measurable if and only if the set $I \setminus E$ is measurable. Furthermore,*

$$m(E) + m(I \setminus E) = m(I),$$

provided that one of the sets E and $I \setminus E$ is measurable.

Proof. By Theorem 2.15,

$$m^*(E) + m_*(I \setminus E) = m(I).$$

By replacing the set E with the set $I \setminus E$ in the above equality, we obtain

$$m_*(E) + m^*(I \setminus E) = m(I).$$

Therefore,

$$m^*(E) - m_*(E) = m^*(I \setminus E) - m_*(I \setminus E).$$

Both claims of the theorem follow immediately from the displayed equalities. □

The property of the Lebesgue measure m known as its *countable additivity* or *σ-additivity* is the result of the next theorem.

Theorem 2.18. *Let a bounded set E be the union of a finite or countable family of pairwise disjoint measurable sets,*

$$E = \bigcup_{i \in J} E_i.$$

Then the set E is measurable and

$$m(E) = \sum_{i \in J} m(E_i).$$

Proof. By Theorem 2.14, the family $\{m_*(E_i)\}_{i \in J}$ is summable and

$$m_*(E) \geq \sum_{i \in J} m_*(E_i) = \sum_{i \in J} m(E_i),$$

and by Theorem 2.13,

$$m^*(E) \leq \sum_{i \in J} m^*(E_i) = \sum_{i \in J} m(E_i).$$

Because $m_*(E) \leq m^*(E)$ (cf. Theorem 2.11), we have

$$\sum_{i \in J} m(E_i) \leq m_*(E) \leq m^*(E) \leq \sum_{i \in J} m(E_i),$$

and the result follows. □

The next two theorems assert that the set of measurable sets of real numbers is closed under finite unions and intersections.

Theorem 2.19. *The union of a finite family of measurable sets is measurable.*

Proof. Let $E = \bigcup_{i \in J} E_i$, where J is a finite set of cardinality n and the sets E_i, $i \in J$ are measurable.

For a given $i \in J$ and an arbitrary $\varepsilon > 0$, there is a bounded open set $G_i \supseteq E_i$ and a closed set $F_i \subseteq E_i$ such that (cf. Exercises 2.13 and 2.14)

$$m(E_i) = m^*(E_i) > m(G_i) - \frac{\varepsilon}{2n}$$

and

$$m(E_i) = m_*(E_i) < m(F_i) + \frac{\varepsilon}{2n},$$

that is,

$$m(G_i) - m(F_i) < \frac{\varepsilon}{n}, \qquad \text{for all } i \in J. \tag{2.5}$$

The sets F_i and $G_i \setminus F_i$ are disjoint and their union is the set G_i. These sets are measurable because F_i is closed and $G_i \setminus F_i = G_i \cap \complement F_i$ is open. Hence, by Theorem 2.18,

$$m(G_i \setminus F_i) = m(G_i) - m(F_i). \tag{2.6}$$

Let $F = \bigcup_{i \in J} F_i$ and $G = \bigcup_{i \in J} G_i$. Clearly, F is a closed set, G is an open set, and both sets are bounded. Arguing as in the previous paragraph, we obtain

$$m(G \setminus F) = m(G) - m(F). \tag{2.7}$$

Because

$$G \setminus F = \left(\bigcup_{i \in J} G_i \right) \setminus F = \bigcup_{i \in J} (G_i \setminus F) \subseteq \bigcup_{i \in J} (G_i \setminus F_i),$$

where all sets on the right and left sides are open, we have by Theorems 2.2 and 2.4

$$m(G \setminus F) \leq \sum_{i \in J} m(G_i \setminus F_i).$$

Because $F \subseteq E \subseteq G$, we have

$$m(F) \leq m_*(E) \leq m^*(E) \leq m(G).$$

Therefore, by (2.5)–(2.7),

$$m^*(E) \quad m_*(E) \le m(G) \quad m(F) = m(G \setminus F)$$
$$\le \sum_{i \in J} m(G_i \setminus F_i) = \sum_{i \in J}[m(G_i) - m(F_i)] < \varepsilon.$$

The desired result follows, because $\varepsilon > 0$ is arbitrary small. · $\qquad\square$

Theorem 2.20. *The intersection of a finite family of measurable sets is measurable.*

Proof. Let $E = \bigcap_{i \in J} E_i$, where J is a finite set and the sets E_i, $i \in J$, are measurable, and let I be an open interval containing the set E,

$$E = \bigcap_{i \in J} E_i \subseteq I.$$

We have (cf. Exercise 1.5b)

$$I \setminus \left(\bigcap_{i \in J} E_i \right) = \bigcup_{i \in J}(I \setminus E_i).$$

By Theorems 2.17 and 2.19, the set on the right side is measurable. Therefore, the set $I \setminus (\bigcap_{i \in J} E_i)$ is measurable. By Theorem 2.17, the set $\bigcap_{i \in J} E_i$ is measurable. $\qquad\square$

More algebraic properties (with respect to set theoretic operations) of the set of measurable sets of real numbers are established in the following theorem.

Theorem 2.21. *Let E_1 and E_2 be two measurable sets. Then*

(i) *The difference $E_1 \setminus E_2$ is measurable.*
(ii) *The symmetric difference $E_1 \triangle E_2$ is measurable.*
     In addition:
(iii) *If $E_2 \subseteq E_1$ and $E = E_1 \setminus E_2$, then $m(E) = m(E_1) - m(E_2)$.*

Proof.

(i) Let I be an open interval containing the union $E_1 \cup E_2$. We have (cf. Exercise 1.3a)
$$E_1 \setminus E_2 = E_1 \cap (I \setminus E_2).$$

By Theorems 2.17 and 2.20, the set $E_1 \setminus E_2$ is measurable.
(ii) Since
$$E_1 \triangle E_2 = (E_1 \setminus E_2) \cup (E_2 \setminus E_1)$$

and the sets $E_1 \setminus E_2$ and $E_2 \setminus E_1$ are disjoint, the set $E_1 \triangle E_2$ is measurable (cf. Theorem 2.18).

(iii) By Theorem 2.18, we have

$$m(E_1) = m(E_2) + m(E),$$

because sets E_2 and E are disjoint and their union is the set E_1. $\square$

The set of measurable sets of real numbers is also closed under countable unions and intersections. To prove these assertions we need a lemma.

Lemma 2.7. *Let* (E_n) *be a sequence of sets and* $E = \bigcup_{i=1}^{\infty} E_i$. *Then the sets*

$$A_1 = E_1, A_2 = E_2 \setminus A_1, \ldots, A_n = E_n \setminus \bigcup_{i=1}^{n-1} A_i, \ldots$$

are pairwise disjoint and their union is the set E,

$$\bigcup_{i=1}^{\infty} A_i = E.$$

Proof. For any $m < n$, we have

$$A_m \cap A_n = A_m \cap \left(E_n \setminus \bigcup_{i=1}^{n-1} A_i \right) = A_m \cap \left(\bigcap_{i=1}^{n-1} (E_n \setminus A_i) \right) = \varnothing,$$

because $A_m \cap (E_n \setminus A_m) = \varnothing$. Hence the sets A_n are pairwise disjoint.
Clearly,

$$\bigcup_{i=1}^{\infty} A_i \subseteq \bigcup_{i=1}^{\infty} E_i = E.$$

Conversely, for a given $x \in E$, let n be the least index i such that $x \in E_i$. Then $x \in A_n \subseteq \bigcup_{i=1}^{\infty} A_i$, and the result follows. $\square$

Theorem 2.22. *Let a bounded set* E *be the union of a countable family of measurable sets* $\{E_i\}_{i \in J}$,

$$E = \bigcup_{i \in J} E_i.$$

Then the set E *is measurable.*

Proof. We may assume that $J = \mathbb{N}$. Let (A_n) be the sequence of sets from Lemma 2.7. A straightforward inductive argument using Theorems 2.19 and 2.21 shows that the sets A_n's are measurable. By Lemma 2.7 and Theorem 2.18, the set E is measurable. $\square$

Theorem 2.23. *The intersection of a countable family of measurable sets is measurable.*

Proof. Let $E = \bigcap_{i=1}^{\infty} E_i$, where E_i's are measurable sets, and let I be an open interval containing the set E,

$$E = \bigcap_{i \in J} E_i \subseteq I.$$

We have (cf. Exercise 1.5b)

$$I \setminus \left(\bigcap_{i \in J} E_i \right) = \bigcup_{i \in J} (I \setminus E_i).$$

By Theorems 2.17 and 2.22, the set on the right side is measurable. Therefore, the set $I \setminus (\bigcap_{i \in J} E_i)$ is measurable. By Theorem 2.17, the set $\bigcap_{i \in J} E_i$ is measurable. $\square$

The last two theorems of this section establish "continuity" properties of Lebesgue's measure.

Theorem 2.24. *Let (E_n) be a sequence of measurable sets such that*

$$E_1 \subseteq E_2 \subseteq \cdots \subseteq E_n \subseteq \cdots$$

and the set $E = \bigcup_{i=1}^{\infty} E_i$ is bounded. Then

$$m(E) = \lim m(E_n).$$

Proof. By Theorem 2.22, the set E is measurable. It is clear that

$$E = E_1 \cup (E_2 \setminus E_1) \cup \cdots \cup (E_{i+1} \setminus E_i) \cup \cdots,$$

where the sets on the right side are pairwise disjoint. Hence, by Theorems 2.18 and 2.21(iii),

$$m(E) = m(E_1) + \sum_{i=1}^{\infty} [m(E_{i+1}) - m(E_i)].$$

The partial sum of the series on the right side is

$$m(E_1) + \sum_{i=1}^{n-1} [m(E_{i+1}) - m(E_i)] = m(E_n).$$

Therefore, $m(E) = \lim_{n \to \infty} m(E_n)$. $\square$

Theorem 2.25. *Let (E_n) be a sequence of measurable sets such that*

$$E_1 \supseteq E_2 \supseteq \cdots \supseteq E_n \supseteq \cdots$$

and let $E = \bigcap_{i=1}^{\infty} E_i$. Then

$$m(E) = \lim m(E_n).$$

Proof. Let I be an open interval containing the set E_1. Then

$$(I \setminus E_1) \subseteq (I \setminus E_2) \subseteq \cdots \subseteq (I \setminus E_n) \subseteq \cdots$$

and

$$I \setminus E = \bigcup_{i=1}^{\infty} (I \setminus E_i).$$

By Theorem 2.24,

$$m(I \setminus E) = \lim_{n \to \infty} m(I \setminus E_n),$$

or equivalently,

$$m(I) - m(E) = \lim_{n \to \infty} [m(I) - m(E_n)].$$

The result follows. $\qquad\qquad\qquad\qquad\qquad\qquad\qquad\qquad\qquad\square$

2.5 Translation Invariance of Measure

For any given real number $a \in \mathbb{R}$ the transformation $\varphi_a : \mathbb{R} \to \mathbb{R}$ given by $\varphi_a(x) = x + a$ is said to be a *translation* of $\mathbb{R}$. The image of a set E under translation φ_a will be denoted by $E + a$, so

$$E + a = \{x + a : x \in E\}.$$

The following theorem lists properties of translations that can be readily verified (cf. Exercise 2.31):

Theorem 2.26. *Let φ_a be a translation of $\mathbb{R}$. Then*

(i) φ_a *is a continuous bijection from $\mathbb{R}$ onto $\mathbb{R}$.*
(ii) *The image of an open set under φ_a is an open set.*
(iii) *Let G be an open set. The component intervals of $G + a$ are exactly the images of the component intervals of the set G under translation φ_a.*

The goal of this section is to establish the following result.

Theorem 2.27. *The image of a measurable set E under the translation φ_a is measurable and*

$$m(E + a) = m(E).$$

In words: the measurability property of sets is invariant under translations and the measure of a measurable set is translation invariant.

We give the proof of Theorem 2.27 as a sequence of lemmas.

Lemma 2.8. *Let I be a bounded open interval. Then*

$$m(I + a) = m(I),$$

for any translation φ_a.

Proof. Clearly, the translate of a bounded open interval is an open interval of the same length. □

Lemma 2.9. *Let G be a bounded open set and φ_a be a translation of $\mathbb{R}$. Then the set $G + a$ is measurable and*

$$m(G + a) = m(G).$$

Proof. It is clear that the set $G + a$ is bounded. It is measurable by Theorem 2.26(ii). Let $\{I\}_{i \in J}$ be the family of component intervals of G. By the previous lemma,

$$m(I_i + a) = m(I_i), \qquad i \in J.$$

Hence, by Theorem 2.26(iii),

$$m(G + a) = \sum_{i \in J} m(I_i + a) = \sum_{i \in J} m(I_i) = m(G),$$

and the result follows. □

Lemma 2.10. *Let E be a bounded set and φ_a be a translation of $\mathbb{R}$. Then*

(i) $m^*(E + a) = m^*(E)$,
(ii) $m_*(E + a) = m_*(E)$.

Proof.

(i) For a given $\varepsilon > 0$, let G be an open set containing E such that

$$m(G) < m^*(E) + \varepsilon$$

(cf. Exercise 2.13). Inasmuch as $E + a \subseteq G + a$, we have, by the previous lemma,

$$m^*(E + a) \leq m^*(G + a) = m(G + a) = m(G) < m^*(E) + \varepsilon.$$

Because ε is an arbitrary positive number,

$$m^*(E + a) \leq m^*(E), \qquad \text{for any } a \in \mathbb{R}.$$

By this inequality (replacing E by $E + a$ and a by $-a$),

$$m^*(E) = m^*((E + a) - a) \leq m^*(E + a).$$

Hence, $m^*(E + a) = m^*(E)$.

(ii) Let I be a bounded open interval containing the set E. Then $E + a \subset I + a$ and $(I \setminus E) + a = (I + a) \setminus (E + a)$ (cf. Exercise 1.12d). By Theorem 2.15,

$$m^*((I + a) \setminus (E + a)) + m_*(E + a) = m(I + a) = m(I).$$

By the result of part (i),

$$m^*((I + a) \setminus (E + a)) = m^*((I \setminus E) + a) = m^*(I \setminus E).$$

From the last two displayed equalities and Theorem 2.15, we obtain

$$m_*(E + a) = m(I) - m^*((I + a) \setminus (E + a))$$
$$= m(I) - m^*(I \setminus E) = m_*(E),$$

which is the desired result.   □

Let E be a measurable set. By the previous lemma, we have

$$m^*(\varphi_a(E)) = m^*(E) = m_*(E) = m_*(\varphi_a(E))$$

for any translation φ_a. It follows that $\varphi_a(E)$ is a measurable set which has the same measure as E,

$$m(\varphi_a(E)) = m(E).$$

This completes the proof of Theorem 2.27.

2.6 The Class of Measurable Sets

The class of all measurable sets includes all open and all closed bounded sets. By Theorems 2.22 and 2.23, bounded countable unions of closed sets and countable intersections of bounded open sets are measurable.

Any bounded countable set E is measurable and its measure is zero. Indeed, E is a countable union of the family of its singletons. Since every singleton is measurable with measure zero, the claim follows from Theorem 2.18. The Cantor set $\mathbf{C}$ (cf. Example 1.2) shows that the converse is false (cf. Exercise 2.4).

By Exercise 2.17, any subset of the Cantor set $\mathbf{C}$ is measurable. Inasmuch as the cardinality of $\mathbf{C}$ is the same as the cardinality of the set of all real numbers (cf. Example 1.2), we conclude that the cardinality of the set of all measurable subsets of $\mathbb{R}$ is the same as the cardinality of all subsets of $\mathbb{R}$. In words, there are as many measurable sets as arbitrary sets of real numbers.

A collection of subsets of a given set E is called a σ *algebra* provided it contains E and is closed with respect to the formation of relative complements and countable unions. By the results of Sect. 2.4, the set of all measurable subsets of a measurable set E is a σ-algebra.

Are there bounded nonmeasurable sets? We conclude this section by constructing an example of such a set.

We say that two real numbers are (rationally) equivalent if their difference is a rational number. The reader should verify that this relation is indeed an equivalence relation (cf. Exercise 2.32a). Since the set of rational numbers is dense in $\mathbb{R}$, each equivalence class of the relation intersects the open interval $(0,1)$ (cf. Exercise 2.32b). From each equivalence class, we select precisely one number in $(0,1)$ and call the resulting set $\mathcal{N}$.

We will show that $\mathcal{N}$ is a nonmeasurable set. First, we establish two properties of the set $\mathcal{N}$.

Lemma 2.11. *If p and q are two distinct rational numbers, then*

$$(\mathcal{N} + p) \cap (\mathcal{N} + q) = \varnothing.$$

Proof. Suppose $x \in (\mathcal{N} + p) \cap (\mathcal{N} + q)$. Then

$$x = \alpha + p = \beta + q, \qquad \text{for some } \alpha, \beta \in \mathcal{N}.$$

Hence, $\alpha - \beta = q - p$ is a rational number. Therefore, the numbers α and β are distinct and belong to the same equivalence class. This contradicts the definition of the set $\mathcal{N}$. $\qquad\square$

Lemma 2.12.
$$(0,1) \subseteq \bigcup_{p \in \mathbb{Q} \cap (-1,1)} (\mathcal{N} + p) \subseteq (-1, 2).$$

Proof. For a given $x \in (0,1)$, let y be a unique element in $\mathcal{N}$ that is rationally equivalent to x, and let $p = x - y$. Since both x and y belong to the interval $(0,1)$, their difference p belongs to the interval $(-1,1)$. Hence, $x \in \mathcal{N} + p$ for $p \in (-1,1)$. This proves the first inclusion.

The second inclusion holds because $\mathcal{N}$ is a subset of $(0,1)$. $\qquad\square$

Theorem 2.28. *The set $\mathcal{N}$ is not measurable.*

Proof. Suppose $\mathcal{N}$ is measurable and let $\gamma = m(\mathcal{N})$.

By Lemma 2.11, the sets $\mathcal{N} + p$, $p \in \mathbb{Q} \cap (-1,1)$, are pairwise disjoint and, by Theorem 2.27, are measurable with the same measure $m(\mathcal{N} + p) = \gamma$. By Lemma 2.12, the set

$$A = \bigcup_{p \in \mathbb{Q} \cap (-1,1)} (\mathcal{N} + p)$$

is bounded. Therefore, by Theorem 2.18, the set A is measurable and

$$m(A) = \gamma + \gamma + \cdots,$$

which is possible only if $m(A) = \gamma = 0$. On the other hand, by Lemma 2.12, $(0,1) \subseteq A$, which implies $1 \leq m(A) = 0$. This contradiction completes the proof. $\qquad\qquad\square$

By Theorem 2.18, the measure of a bounded set enjoys the countable additivity property:

$$m\left(\bigcup_{i \in J} E_i\right) = \sum_{i \in J} m(E_i), \qquad J \text{ is a countable set,}$$

provided that the sets E_i's are pairwise disjoint measurable sets with a bounded union. The construction from Theorem 2.28 shows that the outer measure is not even finitely additive. Indeed, by Theorems 2.12 and 2.13 and by Lemma 2.10, we have

$$1 = m^*(0,1) \leq m^*(A) \leq \sum_{p \in \mathbb{Q} \cap (-1,1)} m^*(\mathcal{N} + p) = \sum_{p \in \mathbb{Q} \cap (-1,1)} m^*(\mathcal{N}).$$

It follows that $m^*(\mathcal{N}) > 0$. Let n be a natural number such that $m^*(\mathcal{N}) > 1/n$ and let J be a finite subset of $\mathbb{Q} \cap (-1,1)$ of cardinality $3n$. If the outer measure m^* was finitely additive, we would have

$$m^*\left(\bigcup_{p \in J} (\mathcal{N} + p)\right) = \sum_{p \in J} m^*(\mathcal{N} + p) = \sum_{p \in J} m^*(\mathcal{N}) > 3n\frac{1}{n} = 3,$$

which contradicts

$$\bigcup_{p \in J} (\mathcal{N} + p) \subseteq \bigcup_{p \in \mathbb{Q} \cap (-1,1)} (\mathcal{N} + p) \subseteq (-1,2).$$

Hence, m^* is not finitely additive.

2.7 Lebesgue Measurable Functions

Theorem 2.29. *Let E be a measurable set and f be a real-valued function $f : E \to \mathbb{R}$. The following statements are equivalent:*

(i) *For each $c \in \mathbb{R}$, the set $\{x \in E : f(x) > c\}$ is measurable.*
(ii) *For each $c \in \mathbb{R}$, the set $\{x \in E : f(x) \geq c\}$ is measurable.*

(iii) *For each $c \in \mathbb{R}$, the set $\{x \in E : f(x) < c\}$ is measurable.*
(iv) *For each $c \in \mathbb{R}$, the set $\{x \in E : f(x) \leq c\}$ is measurable.*

Each of these properties implies that the following sets are measurable:

(a) *For each $c \in \mathbb{R}$, the set $\{x \in E : f(x) = c\}$ is measurable.*
(b) *For any real numbers $c < d$, the set $\{x \in E : c \leq f(x) < d\}$ is measurable.*

Proof. Because

$$\{x \in E : f(x) \leq c\} = E \setminus \{x \in E : f(x) > c\}$$

and

$$\{x \in E : f(x) \geq c\} = E \setminus \{x \in E : f(x) < c\},$$

(i) and (iv) are equivalent, as are (ii) and (iii). Now, (i) implies (ii), because

$$\{x \in E : f(x) \geq c\} = \bigcap_{k=1}^{\infty} \{x \in E : f(x) > c - \tfrac{1}{k}\}$$

and the right side is measurable by Theorem 2.23. Similarly, (ii) implies (i), because

$$\{x \in E : f(x) > c\} = \bigcup_{k=1}^{\infty} \{x \in E : f(x) \geq c + \tfrac{1}{k}\}$$

and the right side is measurable by Theorem 2.22. It follows that statements (i)–(iv) are equivalent. Assuming that one, and hence all, of them holds, we conclude that the set

$$\{x \in E : f(x) = c\} = \{x \in E : f(x) \geq c\} \cap \{x \in E : f(x) \leq c\}$$

is measurable. Similarly, the set

$$\{x \in E : c \leq f(x) < d\} = \{x \in E : f(x) \geq c\} \cap \{x \in E : f(x) < d\}$$

is measurable. □

Note that condition (a) of the theorem does not imply any of the conditions (i)–(iv) (cf. Exercise 2.41), whereas (b) is equivalent to any of these conditions (cf. Exercise 2.42).

Definition 2.6. *A real-valued function on a set E is said to be* measurable *provided its domain E is measurable and it satisfies one of the four statements (i)–(iv) of Theorem 2.29.*

Theorem 2.30. *If f and g are measurable functions on a set E and k is an arbitrary constant, then kf, f^2, $f + g$, and fg are measurable functions on E.*

Proof. If $k = 0$, then kf is a constant function, and hence is measurable (cf. Exercise 2.36). If $k > 0$, then

$$\{x \in E : kf(x) > c\} = \{x \in E : f(x) > c/k\},$$

and if $k < 0$, then

$$\{x \in E : kf(x) > c\} = \{x \in E : f(x) < c/k\},$$

so kf is measurable for any k.

If $c < 0$, then

$$\{x \in E : f^2(x) > c\} = E,$$

and if $c \geq 0$, then

$$\{x \in E : f^2(x) > c\} = \{x \in E : f(x) < -\sqrt{c}\} \cup \{x \in E : f(x) > \sqrt{c}\}.$$

Therefore, f^2 is measurable on E.

Now we show that the function $f + g$ is measurable. For $q \in \mathbb{Q}$ and a given $c \in \mathbb{R}$, let us consider sets

$$A_q = \{x \in E : f(x) < q\} \quad \text{and} \quad B_q = \{x \in E : g(x) < c - q\}.$$

Because functions f and g are measurable, the sets A_q and B_q are measurable for every $q \in \mathbb{Q}$. We have

$$\{x \in E : f(x) + g(x) < c\} = \bigcup_{q \in \mathbb{Q}} A_q \cap B_q. \tag{2.8}$$

Indeed, if $f(x) + g(x) < c$ for some $x \in E$, then $f(x) < c - g(x)$. Then there is a rational number q such that $f(x) < q < c - g(x)$. It follows that $x \in A_q \cap B_q$. On the other hand, if $x \in \bigcup_{q \in \mathbb{Q}} A_q \cap B_q$, then there is $q \in \mathbb{Q}$ such that $x \in A_q$ and $x \in B_q$, so $f(x) + g(x) < c$. Thus, (2.8) holds. The function $f + g$ is measurable, since the right side of (2.8) is a measurable set.

Inasmuch as

$$fg = \tfrac{1}{4}[(f + g)^2 - (f - g)^2],$$

the function fg is measurable. $\qquad\square$

We gave a detailed proof of (2.8) because this kind of construction is quite typical (cf. proof of Theorem 4.6).

For a finite set $\{f_1, \ldots, f_n\}$ of measurable functions with common domain E, we define the function $\min\{f_1, \ldots, f_n\}$ by

$$\min\{f_1, \ldots, f_n\}(x) = \min\{f_1(x), \ldots, f_n(x)\}, \qquad \text{for } x \in E.$$

The function $\max\{f_1, \ldots, f_n\}$ is defined the same way.

Theorem 2.31. *For a finite family $\{f_i\}_{i=1}^n$ of measurable functions on a set E, the functions $\min\{f_1, \ldots, f_n\}$ and $\max\{f_1, \ldots, f_n\}$ are also measurable.*

Proof. For any given $c \in \mathbb{R}$, we have

$$\{x \in E : \min\{f_1, \ldots, f_n\}(x) > c\} = \bigcap_{i=1}^{n} \{x \in E : f_i(x) > c\},$$

so this set is measurable as a finite intersection of measurable sets. Hence, the function $\min\{f_1, \ldots, f_n\}$ is measurable. A similar argument shows that the function $\max\{f_1, \ldots, f_n\}$ also is measurable. □

Corollary 2.1. *If f is a measurable function on a set E, then $|f|$ is also measurable on E.*

Proof. The claim follows immediately from

$$|f|(x) = \max\{f(x), -f(x)\}$$

and Theorems 2.30 and 2.31. □

2.8 Sequences of Measurable Functions

For notation convenience, we often write "$f \leq g$ on E" for "$f(x) \leq g(x)$ for all $x \in E$" (similarly, "$f < g$ on E" for "$f(x) < g(x)$ for all $x \in E$") and use the same convention for equalities.

Definition 2.7. *Let (f_n) be a sequence of functions with a common domain E, f be a function on E, and A be a subset of E. We say that*

(i) *The sequence (f_n) converges to f pointwise on A if*

$$\lim f_n(x) = f(x) \quad \text{for all } x \in A.$$

(ii) *The sequence (f_n) converges to f pointwise almost everywhere on A provided it converges to f pointwise on $A \setminus B$, where $m(B) = 0$.*

(iii) *The sequence (f_n) converges to f uniformly on A provided for each $\varepsilon > 0$ there is an index N such that*

$$|f - f_n| < \varepsilon, \quad \text{for all } n \geq N.$$

Observe that uniform convergence of a sequence (f_n) implies its pointwise convergence.

In measure theory, we say that a property holds *almost everywhere* (abbreviated a.e.) on a measurable set E provided it holds on $E \setminus E_0$, where E_0 is a subset of E of measure zero. Definition 2.7(ii) accounts for situations when we might have a few values of $x \in E$ at which the sequence $(f_n(x))$ does not converge or converges to a number different from $f(x)$.

Example 2.3. (The Cantor function). Let f be an increasing linear function that maps the interval $[a, b]$ onto the interval $[c, d]$. If m denotes the slope of the graph of f, that is, $m = (d - c)/(b - a)$, then

$$f(x) = mx + (c - ma) \quad \text{on } [a, b].$$

Let us define a new function Tf by

$$(Tf)(x) = \begin{cases} \frac{3}{2}mx - (c - \frac{3}{2}ma), & \text{if } a \le x \le \frac{2}{3}a + \frac{1}{3}b, \\ \frac{1}{2}c + \frac{1}{2}d, & \text{if } \frac{2}{3}a + \frac{1}{3}b < x \le \frac{1}{3}a + \frac{2}{3}b, \\ \frac{3}{2}mx + (d - \frac{3}{2}mb), & \text{if } \frac{1}{3}a + \frac{2}{3}b < x \le b. \end{cases}$$

Graphs of functions $f_0(x) = x$ on $[0, 1]$ and $f_1 = Tf_0$ on the same interval are shown in Fig. 2.5 left.

Now we apply the transformation T to two linear "pieces" of the function f_1 that have slope $3/2$ to obtain function f_2 whose graph is shown in Fig. 2.5 right. Observe that f_2 is again a piecewise linear function on $[0, 1]$ and that the slopes of all non-horizontal "pieces" are $(3/2)^2$.

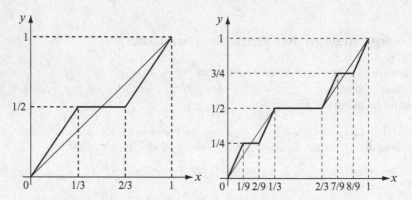

Figure 2.5. Functions $f_1(x)$ *(left)* and $f_2(x)$ *(right)*

By continuing this process, we define a piecewise linear function f_k on $[0, 1]$ that has "pieces" with nonzero slopes over the set C_k (cf. Example 1.2) and assumes constant values $\frac{1}{2^k}, \ldots, \frac{2^k - 1}{2^k}$ on the component intervals of the open set $[0, 1] \setminus C_k$.

It is not difficult to show that the sequence (f_n) converges to a nondecreasing function $c(x)$ that maps the interval $[0, 1]$ onto itself (cf. Exercise 2.48). The function $c(x)$ is called the *Cantor function*.

Inasmuch as each function f_k is piecewise linear, it has the left-hand derivative $D^- f_k$ on $(0, 1]$. We have

$$D^- f_k(x) = \begin{cases} (3/2)^k, & \text{if } x \in C_k \setminus \{0\}, \\ 0, & \text{if } x \in [0, 1] \setminus C_k. \end{cases}$$

Because the measure of the Cantor set is zero, the sequence $(D^- f_n)$ converges to zero almost everywhere. Moreover, $c'(x) = 0$ a.e. on $[0, 1]$.

Theorem 2.32. *Let (f_n) be a sequence of measurable functions on E that converges pointwise to the function f. Then f is measurable.*

Proof. By Theorem 1.5,

$$f(x) = \limsup f_n(x) = \lim g_n(x),$$

where

$$g_n(x) = \sup\{f_k(x) : k \geq n\}.$$

For any given x, the sequence $(f_n(x))$ is bounded (because it is convergent), so the functions g_n's are well defined.

For given $c \in \mathbb{R}$ and $x \in E$,

$$g_n(x) \leq c \quad \text{if and only if} \quad f_k(x) \leq c \text{ for all } k \geq n$$

(cf. Exercise 1.19). Therefore,

$$\{x \in E : g_n(x) \leq c\} = \bigcap_{k \geq n}\{x \in E : f_k(x) \leq c\}.$$

Because the functions f_n's are measurable, the set on the right side is measurable. It follows that functions g_n, $n \in \mathbb{N}$, are measurable.

By Theorem 1.3,

$$f(x) = \inf\{g_n(x) : n \in \mathbb{N}\},$$

inasmuch as $(g_n(x))$ is clearly a decreasing sequence for any $x \in E$. We have

$$f(x) \geq c \quad \text{if and only if} \quad g_n(x) \geq c \text{ for all } n \in \mathbb{N}.$$

for any given $c \in \mathbb{R}$ and $x \in E$ (cf. Exercise 1.19). Therefore,

$$\{x \in E : f(x) \geq c\} = \bigcap_{n \in \mathbb{N}}\{x \in E : g_n(x) \geq c\},$$

that is, f is a measurable function on E. □

Theorem 2.33. *Let (f_n) be a sequence of measurable functions on E that converges pointwise a.e. to the function f. Then f is measurable.*

Proof. Let us denote by E_0 the set of points $x \in E$ for which

$$\lim f_n(x) = f(x)$$

does not hold. The function f is measurable on E_0 since $m(E_0) = 0$ (cf. Exercise 2.35). By Theorem 2.32, f is measurable on $E \setminus E_0$. Clearly,

$$\{x \in E : f(x) > c\} = \{x \in E_0 : f(x) > c\} \cup \{x \in E \setminus E_0 : f(x) > c\}.$$

Thus f is measurable on E. □

We conclude this chapter by proving a remarkable result known as Egorov's Theorem. Informally, it states that "every convergent sequence of measurable functions is nearly uniformly convergent."

Theorem 2.34. (Egorov's Theorem) *Let (f_n) be a sequence of measurable functions on E that converges pointwise to the function f. Then for each $\delta > 0$, there is a measurable set $E_\delta \subseteq E$ such that $m(E_\delta) < \delta$ and (f_n) converges uniformly to f on $E \setminus E_\delta$.*

Proof. By Theorem 2.32, f is a measurable function. For a given $\sigma > 0$ we define two sequences of measurable sets

$$A_n(\sigma) = \{x \in E : |f_n(x) - f(x)| \geq \sigma\}$$

and

$$B_n(\sigma) = \bigcup_{k \geq n} A_k(\sigma).$$

It is clear that $(B_n(\sigma))$ is a decreasing family of sets, that is,

$$B_1(\sigma) \supseteq B_2(\sigma) \supseteq \cdots \supseteq B_n(\sigma) \supseteq \cdots .$$

Moreover,

$$\bigcap_{n=1}^{\infty} B_n(\sigma) = \varnothing.$$

Indeed, since $(f_n(x))$ converges to $f(x)$ for a given $x \in E$, there is an index N such that $|f_n(x) - f(x)| < \sigma$ for all $n \geq N$, that is, $x \notin A_n(\sigma)$ for $n \geq N$. It follows that for any given $x \in E$, there is N such that $x \notin B_N(\sigma)$. Hence, the intersection $\bigcap_{n=1}^{\infty} B_n(\sigma)$ is empty.

By Theorem 2.25, $\lim m(B_n(\sigma)) = 0$. Therefore, for each $k \in \mathbb{N}$, there is n_k such that

$$m(B_{n_k}(1/k)) < \frac{\delta}{2^k},$$

where $\delta > 0$ is a given number. We define

$$E_\delta = \bigcup_{k=1}^{\infty} B_{n_k}(1/k).$$

Then

$$m(E_\delta) \leq \sum_{k=1}^{\infty} m(B_{n_k}(1/k)) < \sum_{k=1}^{\infty} \frac{\delta}{2^k} = \delta.$$

Let $c > 0$ be a given number and let us choose k so $1/k < \varepsilon$. For every $x \in E \setminus E_\delta$, we have $x \notin B_{n_k}(1/k)$ and hence

$$x \notin \bigcup_{n \geq n_k} A_n(1/k),$$

which implies

$$|f_n(x) - f(x)| < 1/k < \varepsilon \quad \text{for } n \geq n_k,$$

for all $x \in E \setminus E_\delta$. Hence, (f_n) converges uniformly to f over the set $E \setminus E_\delta$.

$\square$

Notes

The following passage from Lebesgue's book *Leçons sur l'intégration et la recherche des fonctions primitives* is worth quoting in its entirety:

Nous nous proposons d'attacher à chaque ensemble **E** *borné, formé de points de ox, un nombre positif ou nul, $m(\mathbf{E})$, que nous appelons la mesure de* **E** *et qui satisfait aux conditions suivantes:*

1. *Deux ensembles égaux ont même mesure*
2. *L'ensemble somme d'un nombre fini ou d'une infinité dénombrable d'ensembles, sans point commun deux à deux, a pour mesure la somme des mesures*
3. *La mesure de l'ensemble de tous les points de $(0,1)$ est 1*

[Leb28, p. 110]

Here is its translation:

"We propose to attach to each bounded set **E**, made up of points of the x-axis, a nonnegative number $m(\mathbf{E})$, that we call the measure of **E** and that satisfies the following conditions:

1. Two equal sets have same measure
2. The measure of the sum of a finite or countably infinite number of sets, without common points between any two sets, is the sum of the measures
3. The measure of the set made up of all points of $(0,1)$ is 1"

The reader can recognize Lebesgue's property (1) as the translation invariance property (cf. Sect. 2.5). Property (2) is the countable additivity of Lebesgue's measure (cf. Theorem 2.18).

In his book [Leb28], Lebesgue proceeds by introducing the outer measure as we present it in this chapter. Then he defines the inner measure of a set E which is a subset of an interval $I = (a, b)$ as

$$m_*(E) = m(I) - m^*(I \setminus E)$$

(cf. Theorem 2.15) and a measurable set as a set such that its outer and inner measures are equal. Our exposition of this subject is similar to the one found in the classical text [Nat55].

Egorov's Theorem (Theorem 2.34) is a nontrivial result in this chapter. This theorem is of great importance in the studies of convergence of integrals in Chap. 3.

Note that the result of Egorov's Theorem cannot be strengthened to include, in some sense, the case of $\delta = 0$ (cf. Exercise 2.50). In this connection, see Theorem A.8.

Exercises

2.1. Show that the set $\bigcup_{k=1}^{\infty}(\frac{1}{k+1}, \frac{1}{k})$ is open and find its measure.

2.2. Show that the set $\bigcup_{k=1}^{\infty}(\frac{1}{2k}, \frac{1}{2k-1})$ is open and find its measure. (Hint: cf. Exercise 1.53.)

2.3. Let $X \subseteq Y \subseteq Z$ be three sets. Show that the sets $Z \setminus Y$ and $Y \setminus X$ are disjoint, and

$$Z \setminus X = (Z \setminus Y) \cup (Y \setminus X),$$

2.4. Show that measure of the Cantor set $\mathbf{C}$ is zero.

2.5. Let us define $\mathbf{C}(n)$ as the set that remains after removing from $[0, 1]$ an open interval of length $1/n$ centered at $1/2$, then an open interval of length $1/n^2$ from the center of each of the two remaining intervals, then open intervals of length $1/n^3$ from the centers of each of the remaining four intervals, and so on. Note that $\mathbf{C}(3) = \mathbf{C}$, the Cantor set.

Show that $\mathbf{C}(n)$ is a closed set and

$$m(\mathbf{C}(n)) = \frac{n-3}{n-2}.$$

2.6. For sets $A \subseteq B \subseteq C$ show that

$$C = B \cup (C \setminus A).$$

2.7. Let $I = (a, b)$ be an open interval. Show that for every positive $\delta < b - a$ there is a closed interval $[a', b'] \subseteq (a, b)$ such that

$$m([a', b']) > m(I) - \delta.$$

2.8. Prove the second inequality in Theorem 2.12.

2.9. Let E and S be bounded sets. Prove that if $m^*(E) = 0$ then $m^*(E \cup S) = m^*(S)$.

2.10. Find the outer and inner measures of the following sets:

(a) $\mathbb{Q} \cap [0, 1]$.
(b) $[0, 1] \setminus \mathbb{Q}$.

2.11. Show that the outer measure of a singleton is zero. Deduce that the set $[0, 1]$ is not countable.

2.12. Let A and B be bounded sets for which there is an $\varepsilon > 0$ such that $|a - b| \geq \varepsilon$ for all $a \in A$, $b \in B$. Prove that

$$m^*(A \cup B) = m^*(A) + m^*(B).$$

2.13. Prove that for any bounded set E and any $\varepsilon > 0$, there is an open set $G \supseteq E$ such that
$$m(G) < m^*(E) + \varepsilon.$$

2.14. Prove that for any bounded set E and any $\varepsilon > 0$, there exists a closed set $F \subseteq E$ such that
$$m(F) > m_*(E) - \varepsilon.$$

2.15. Prove that for any bounded set E, there is a bounded set A that is a countable intersection of open sets for which $E \subseteq A$ and

$$m^*(E) = m^*(A).$$

2.16. Prove that for any bounded set E, there exists a set B that is a countable union of closed sets for which $E \supseteq B$ and

$$m_*(E) = m_*(B).$$

2.17. Let E be a set of measure zero. Prove that any subset of E is measurable and its measure is zero.

2.18. Show that if E and E' are measurable sets, then

$$m(E \cup E') + m(E \cap E') = m(E) + m(E').$$

2.19. Let E be a bounded set. Show that if there is a measurable subset $E' \subseteq E$ such that $m(E') = m^*(E)$, then E is measurable.

2.20. Prove that a bounded set E is measurable if and only if for every $\varepsilon > 0$ there exists a closed set $F \subseteq E$ such that $m^*(E \setminus F) < \varepsilon$.
(de la Vallée Poussin Criterion).

2.21. Let A and B be two measurable disjoint sets. Show that for any set E

$$m^*[E \cap (A \cup B)] = m^*(E \cap A) + m^*(E \cap B)$$

and

$$m_*[E \cap (A \cup B)] = m_*(E \cap A) + m_*(E \cap B).$$

2.22. Prove that a bounded set E is measurable if and only if for every bounded set A we have

$$m^*(A) = m^*(A \cap E) + m^*(A \setminus E).$$

(Carathéodory Criterion).

2.23. Show that for any bounded set E, the following statements are equivalent:

(a) E is measurable.
(b) Given any $\varepsilon > 0$, there is an open set $G \supseteq E$ such that

$$m^*(G \setminus E) < \varepsilon.$$

(c) Given any $\varepsilon > 0$, there is a closed set $F \subseteq E$ such that

$$m^*(E \setminus F) < \varepsilon.$$

2.24. Show that a set E is measurable if and only if for each $\varepsilon > 0$, there is a closed set F and open set G for which $F \subseteq E \subseteq G$ and $m^*(G \setminus F) < \varepsilon$.

2.25. Let E be a measurable set and $\varepsilon > 0$. Show that E is a union of a finite family of pairwise disjoint measurable sets, each of which has measure at most ε.

2.26. Let E be a measurable set. Show that for each $\varepsilon > 0$ there is a finite family of pairwise disjoint open intervals $\{I_i\}_{i \in J}$ such that

$$m(E \bigtriangleup G) < \varepsilon,$$

where $G = \bigcup_{i \in J} I_i$.

2.27. Show that a bounded set E is measurable if and only if for each open interval $I = (a, b)$,

$$b - a = m^*(I \cap E) + m^*(I \setminus E).$$

2.28. Let $\{E_i\}_{i \in J}$ be a countable family of measurable pairwise disjoint sets. Prove that for any bounded set A

$$m^*\left(A \cap \bigcup_{i \in J} E_i\right) = \sum_{i \in J} m^*(A \cap E_i).$$

2.29. Let E be a measurable set of real numbers. Show that the function

$$f(x) = m(E \cap (-\infty, x])$$

is continuous.

2.30. Let A be a measurable subset of the interval $(0,1)$ such that $m(A) = 1$. Show that $\inf A = 0$ and $\sup A = 1$.

2.31. Prove Theorem 2.26.

2.32. Define a binary relation R on $\mathbb{R}$ by

$$R = \{(x,y) \in \mathbb{R}^2 : x - y \in \mathbb{Q}\}.$$

(a) Show that R is an equivalence relation (cf. Sect. 1.1) on $\mathbb{R}$.
(b) Show that each equivalence class of R has a nonempty intersection with $(0,1)$.

2.33. Define a binary relation R on $\mathbb{R}$ by

$$R = \{(x,y) \in \mathbb{R}^2 : x - y \in \mathbb{R} \setminus \mathbb{Q}\}.$$

Show that R is symmetric but not reflexive and not transitive binary relation on $\mathbb{R}$.

2.34. Show that any measurable set with positive measure contains a nonmeasurable subset.

2.35. Show that any function on a set of measure zero is measurable.

2.36. Show that any constant function on a measurable set is measurable.

2.37. A function f on a closed interval $[a,b]$ is said to be a *step function* if there is a sequence of points

$$c_0 = a < c_1 < c_2 < \cdots < c_n = b$$

such that f is constant on each open interval (c_k, c_{k+1}), $0 \le k < n$. Prove that a step function is measurable.

2.38. If $|f|$ is measurable, does it necessarily follow that f is measurable?

2.39. Prove that a function f on $[a,b]$ is measurable if and only if $f^{-1}(U)$ is measurable for any open set $U \subseteq \mathbb{R}$.

2.40. Prove that any continuous function on $[a,b]$ is measurable.

2.41. Suppose that f is a function on $[a,b]$ such that

$$\{x \in [a,b] : f(x) = c\}$$

is measurable for each number c. Is f necessarily measurable?

2.42. Suppose that f is a function on E such that the set

$$\{x \in E : c \le f(x) < d\}$$

is measurable for any $c < d$. Show that f is measurable.

2.43. If $(f(x))^n$ is measurable for some $n \in \mathbb{N}$, does it necessarily follow that f is measurable?

2.44. Prove that if f is measurable on E and $g = f$ a.e. on E, then g is measurable.

2.45. Let f and g be continuous functions on $[a, b]$. Show that if $f = g$ a.e. on $[a, b]$, then, in fact, $f = g$ on $[a, b]$.

2.46. Show that an increasing function f on the closed interval $[a, b]$ is measurable. (Hint: consider first the strictly increasing function $f(x) + x/n$ for a given $n \in \mathbb{N}$.)

2.47. Let (f_n) be a sequence of measurable functions on E that converges pointwise to f. Show that for any $\varepsilon > 0$, there is a closed set $F \subseteq E$ for which (f_n) converges uniformly on F and $m(E \setminus F) < \varepsilon$.

2.48. Show that the Cantor function $c(x)$ (cf. Example 2.3)

(a) is a continuous nondecreasing function from $[0, 1]$ onto $[0, 1]$.
(b) maps the Cantor set $\mathbf{C}$ onto the interval $[0, 1]$.

2.49. Let (f_n) be a sequence of measurable functions on E. Show that the set A of points at which this sequence converges is measurable.

2.50. Give an example of a piecewise convergent sequence of measurable functions on a measurable set E that does not converge uniformly almost everywhere on E.

3

Lebesgue Integration

We define the Lebesgue integral in three stages. First, we define the integral of a bounded function over a measurable set E by following the original Lebesgue's method. Then, for a nonnegative measurable function f on E, the integral $\int_E f$ is defined as the supremum of the integrals of lower approximations of f by bounded functions, and the function f is called integrable over E if $\int_E f$ is finite. Finally, a general measurable function f is said to be integrable over E if its positive and negative parts f^+ and f^- are integrable over E. Then $\int_E f$ is defined as $\int_E f^+ - \int_E f^-$.

For a fixed measurable set E, the integral $\int_E f$ is a function on the set of integrable functions f on E. We prove that this function is linear and monotone. On the other hand, for a fixed integrable function f on E, the integral $\int_A f$ is a function on measurable subsets A of E. We show that this function enjoys additivity properties.

A distinguished feature of the Lebesgue integral is that it allows for the "passage of the limit under the integral sign," that is, $\lim \int_E f_n = \int_E (\lim f_n)$, under some quite general assumptions about converging sequences of functions. By relying on the power of Egorov's Theorem (Theorem 2.34), we establish such classical results as the bounded convergence theorem (Theorem 3.10), the monotone convergence theorem (Theorem 3.17),and the dominated convergence theorem (Theorem 3.25).

3.1 Integration of Bounded Measurable Functions

We begin our exposition by recalling the concept of a partition from real analysis.

Let I be an arbitrary nontrivial bounded interval with $\inf I = a$, $\sup I = b$ (so $a < b$). A *partition* of I is a set of points

$$P = \{y_0, y_1, \ldots y_n\}, \qquad n > 1,$$

S. Ovchinnikov, *Measure, Integral, Derivative: A Course on Lebesgue's Theory*,
Universitext, DOI 10.1007/978-1-4614-7196-7_3,
© Springer Science+Business Media New York 2013

such that
$$a = y_0 < y_1 < \cdots < y_n = b.$$

The *norm* of a partition $P = \{y_0, y_1, \ldots y_n\}$ is the number

$$\|P\| = \max\{y_i - y_{i-1} : 1 \le i \le n\}.$$

If P and Q are two partitions of (A, B) such that $P \subseteq Q$, then we say that Q is *finer* than P.

Let f be a bounded measurable function on a set E and let (A, B) be an arbitrary open interval containing all values of the function f, that is, $f(E) \subseteq (A, B)$. For a partition $P = \{y_0, y_1, \ldots, y_n\}$ of (A, B), we define subsets $E_k(f, P)$ of E by

$$E_k(f, P) = \{x \in E : y_k \le f(x) < y_{k+1}\}, \quad \text{for } k = 0, 1, \ldots, n-1.$$

Each set $E_k(f, P)$, $0 \le k < n$, is the inverse image of the half-open interval $[y_k, y_{k+1})$ under the mapping $f : E \to \mathbb{R}$. It is easy to verify the following properties of these sets:

1. The sets $E_k(f, P)$ are pairwise disjoint.
2. The sets $E_k(f, P)$ are measurable.
3. $E = \bigcup_{k=0}^{n-1} E_k(f, P)$.
4. $m(E) = \sum_{k=0}^{n-1} m(E_k(f, P))$.

The lower and upper *Lebesgue sums* are defined by

$$s(f, P) = \sum_{k=0}^{n-1} y_k m(E_k(f, P)), \qquad S(f, P) = \sum_{k=0}^{n-1} y_{k+1} m(E_k(f, P)).$$

Note that these sums depend also on the choice of the interval (A, B).
It is evident that $s(f, P) \le S(f, P)$. For $\lambda = \|P\|$, we have

$$\sum_{k=0}^{n-1} (y_{k+1} - y_k) m(E_k(f, P)) \le \lambda m(E).$$

Hence,
$$0 \le S(f, P) - s(f, P) \le \lambda m(E).$$

Clearly, we can make λ as small as we wish by choosing a sufficiently fine partition of (A, B).

Lemma 3.1. *Let $s(f, P)$ and $S(f, P)$ be the Lebesgue sums for a partition $P = \{y_0, y_1, \ldots, y_n\}$ of the interval (A, B). If we add a new point, y', to this set, and compute the Lebesgue sums $s(f, P')$ and $S(f, P')$ for the partition $P' = P \cup \{y'\}$, then we obtain*

$$s(f, P) \le s(f, P'), \quad S(f, P') \le S(f, P).$$

In words, by adding new division points we do not decrease the lower sums and do not increase the upper sums.

Proof. Suppose that $y_k < y' < y_{k+1}$. Each term $y_i m(E_i(f, P'))$ with $i \neq k$ of the new lower sum $s(f, P')$ is also a term of the sum $s(f, P)$. The term $y_k m(E_k(f, P))$ of the sum $s(f, P)$ is replaced by two new terms, $y_k m(E_k')$ and $y' m(E_k'')$, where sets E_k' and E_k'' are inverse images of the half-open intervals $[y_k, y')$ and $[y', y_{k+1})$, respectively. Because $E_k(f, P) = E_k' \cup E_k''$ and $E_k' \cap E_k'' = \varnothing$, we have

$$y_k m(E_k(f, P)) = y_k m(E_k') + y_k m(E_k'') \leq y_k m(E_k') + y' m(E_k'').$$

Hence, $s(f, P) \leq s(f, P')$.

A similar argument shows that $S(f, P') \leq S(f, P)$. □

Lemma 3.2. $s(f, P) \leq S(f, Q)$, *for all partitions* P *and* Q *of the interval* (A, B).

Proof. Let $s(f, P)$, $s(f, Q)$ and $S(f, P)$, $S(f, Q)$ be the lower and upper sums corresponding to two partitions of the interval (A, B). Let us join the points of these two partitions and compute the new lower and upper sums $s(f, P \cup Q)$ and $S(f, P \cup Q)$, respectively. By Lemma 3.1, $s(f, P) \leq s(f, P \cup Q)$ and $S(f, P \cup Q) \leq S(f, Q)$. Inasmuch as $s(f, P \cup Q) \leq S(f, P \cup Q)$, we obtain the desired result, $s(f, P) \leq S(f, Q)$. □

Theorem 3.1. *For a bounded measurable function* $f : E \to \mathbb{R}$, *the supremum and infimum,* $U(f)$ *and* $V(f)$, *of the sets of Lebesgue lower and upper sums, respectively, are well defined and equal,*

$$U(f) = V(f).$$

Note that the supremum and infimum in the theorem are taken over partitions of the interval (A, B).

Proof. By Lemma 3.2, the sets of lower and upper sums are bounded above and below, respectively. Therefore, quantities $U(f)$ and $V(f)$ are well defined. By the same lemma, $U(f) \leq V(f)$. For every partition P of (A, B) we clearly have

$$s(f, P) \leq U(f) \leq V(f) \leq S(f, P).$$

As we noted before, $S(f, P) - s(f, P) \leq \lambda\, m(E)$, where $\lambda = \|P\|$, and therefore

$$0 \leq V(f) - U(f) \leq \lambda\, m(E).$$

Because λ can be made arbitrary small, we obtain $U(f) = V(f)$. □

Definition 3.1. *Let f be a bounded measurable function on a set E. The common value of numbers $U(f)$ and $V(f)$ is called the* Lebesgue integral *of f and is denoted by the symbol*

$$\int_E f.$$

We write $(L)\int_E f$ if it is desirable to distinguish the Lebesgue integral from a different type of integral such as the Riemann integral (cf. Scct. 3.6). If E is a closed interval $[a, b]$, the symbols

$$(L)\int_a^b f, \quad \int_a^b f, \quad \text{and} \quad \int_a^b f(x)\,dx$$

are also used.

We show now that the value of the Lebesgue integral does not depend on the choice of the interval (A, B). Suppose, for instance, that

$$A < f(x) < B < B', \qquad \text{for all } x \in E.$$

Let $\{A, y_1, \ldots, y_{n-1}, B\}$ be a partition of (A, B) defining the lower sum s. Then $\{A, y_1, \ldots, y_{n-1}, B'\}$ is a partition of (A, B') defining the same lower sum s. On the other hand, it is clear that any partition of (A, B') defining the lower sum s defines exactly the same lower sum s when restricted to (A, B). Therefore the sets of lower sums obtained from intervals (A, B) and (A, B') are the same and hence they give the same value for the Lebesgue integral. The same result also holds for the lower end of the interval (A, B).

We conclude that

$$\int_E f = \sup\{s : s \text{ is a Lebesgue lower sum}\} \tag{3.1}$$

$$= \inf\{S : S \text{ is a Lebesgue upper sum}\}, \tag{3.2}$$

where the lower and upper sums are taken for an arbitrary interval (A, B) containing the range of the function f.

3.2 Properties of the Integral

The following theorem establishes an important property of the Lebesgue integral which is used frequently in this chapter.

Theorem 3.2. *Let f be a measurable function on a set E. If*

$$a \leq f(x) \leq b, \qquad \text{for all } x \in E,$$

then

$$a \cdot m(E) \leq \int_E f \leq b \cdot m(E).$$

Proof. For a given $\varepsilon > 0$, we define $A = a - c$ and $B = b + c$. Then

$$A < f(x) < B, \qquad \text{for all } x \in E.$$

Let $P = \{y_0, y_1, \ldots, y_n\}$ be a partition of the interval (A, B). We have

$$A \sum_{k=0}^{n-1} m(E_k(f, P)) \leq \sum_{k=0}^{n-1} y_k m(E_k(f, P)) \leq B \sum_{k=0}^{n-1} m(E_k(f, P)),$$

that is,

$$A \cdot m(E) \leq s(f, P) \leq B \cdot m(E).$$

By (3.1),

$$(a - \varepsilon) m(E) \leq \int_E f \leq (b + \varepsilon) m(E).$$

We obtain the desired result by taking $\varepsilon \to 0$. $\qquad\qquad\qquad\square$

The quantity $\frac{1}{m(E)} \int_E f$ is called the "mean value" of f over the set E, provided that $m(E) \neq 0$. Theorem 3.2 states that the mean value of a bounded measurable function lies between its lower and upper bounds. For this reason Theorem 3.2 is sometimes called the "first law of the mean for integrals."

Lemma 3.3. *Let f be a bounded measurable function on E. If*

$$E = \bigcup_{k=1}^{n} A_k,$$

where the sets A_k are measurable and pairwise disjoint, then

$$\int_E f = \sum_{k=1}^{n} \int_{A_k} f.$$

Proof. We prove the lemma by induction on n. The claim is trivial for $n = 1$.

Suppose that $n > 1$ and let $E' = \bigcup_{k=1}^{n-1} A_k$, $E'' = A_n$. Then, $E = E' \cup E''$ and $E' \cap E'' = \varnothing$.

Let us partition an arbitrary open interval containing the set $f(E)$ by points $y_0 < y_1 < \cdots < y_m$ and consider sets

$$E_i = \{x \in E : y_i \leq f(x) < y_{i+1}\},$$
$$E_i' = \{x \in E' : y_i \leq f(x) < y_{i+1}\},$$
$$E_i'' = \{x \in E'' : y_i \leq f(x) < y_{i+1}\},$$

for $0 \leq i < m$. Evidently, $E_i = E_i' \cup E_i''$ and $E_i' \cap E_i'' = \varnothing$. Therefore,

$$\sum_{i=0}^{m-1} y_i m(E_i) = \sum_{i=0}^{m-1} y_i m(E_i') + \sum_{i=0}^{m-1} y_i m(E_i''),$$

that is,

$$s = s' + s'',$$

where s, s', and s'' are the lower sums for the function f over sets E, E', and E'', respectively, defined by the partition $\{y_0, y_1, \ldots, y_n\}$. By (3.1),

$$s' \leq \int_{E'} f \quad \text{and} \quad s'' \leq \int_{E''} f,$$

so

$$s = s' + s'' \leq \int_{E'} f + \int_{E''} f.$$

By (3.1) again,

$$\int_E f \leq \int_{E'} f + \int_{E''} f.$$

To prove the opposite inequality, we consider the relation between the upper sums

$$S = S' + S''.$$

By applying (3.2), we obtain

$$S = S' + S'' \geq \int_{E'} f + \int_{E''} f,$$

which implies

$$\int_E f \geq \int_{E'} f + \int_{E''} f.$$

Thus

$$\int_E f = \int_{E'} f + \int_{E''} f = \int_{E'} f + \int_{A_n} f.$$

By the induction hypothesis,

$$\int_{E'} f = \sum_{k=1}^{n-1} \int_{A_k} f,$$

which yields the desired result. $\square$

The next theorem extends the result of Lemma 3.3 to countable families of pairwise disjoint sets. The proof illustrates the power of the "first law of the means."

Theorem 3.3. *Let f be a bounded measurable function on E. If*

$$E = \bigcup_{i \in J} A_i,$$

where $\{A_i\}_{i\in J}$ is at most countable family of pairwise disjoint measurable sets,
then

$$\int_E f = \sum_{i\in J} \int_{A_i} f.$$

Proof. By Lemma 3.3, it suffices to consider the case of a countable set J.

The assertion obviously holds for the zero function on E, so we assume
that f is not the zero function. Suppose that $f(E) \subseteq [a,b]$ and let ε be a given
positive number. By Theorem 2.18, the family $\{m(A_i)\}_{i\in J}$ is summable with
the sum

$$m(E) = \sum_{i\in J} m(A_i).$$

Therefore there is a finite set $J_0 \subset J$ such that

$$\left| m(E) - \sum_{i\in J_0} m(A_i) \right| < \frac{\varepsilon}{c},$$

where $c = \max\{|a|, |b|\}$. Note that $c \neq 0$ because we assumed that f is not
the zero function. Let

$$B = \bigcup_{i\in J\setminus J_0} A_i,$$

so

$$E = B \cup \bigcup_{i\in J_0} A_i \quad \text{and} \quad m(B) < \varepsilon/c.$$

By Lemma 3.3,

$$\int_E f = \sum_{i\in J_0} \int_{A_i} f + \int_B f,$$

and, by Theorem 3.2,

$$a \cdot m(B) \leq \int_B f \leq b \cdot m(B).$$

Hence,

$$\left| \int_E f - \sum_{i\in J_0} \int_{A_i} f \right| = \left| \int_B f \right| < \frac{\varepsilon}{c} \cdot \max\{|a|, |b|\} = \varepsilon$$

(cf. Exercise 3.2). It follows that the family $\{\int_{A_i} f\}_{i\in J}$ is summable with the
sum $\int_E f$. $\square$

Two important corollaries follow from Theorem 3.3.

Corollary 3.1. *If bounded measurable functions f and g are equal a.e. on E, then*

$$\int_E f = \int_E g.$$

The proof is left as Exercise 3.3.

Example 3.1. (Dirichlet's function) We define

$$f(x) = \begin{cases} 1, & \text{if } x \text{ is a rational number,} \\ 0, & \text{otherwise,} \end{cases} \qquad x \in [0,1].$$

Because the set of rational numbers in $[0,1]$ is countable, its measure is zero. Thus, f equals the zero function a.e. on $[0,1]$. It follows that f is Lebesgue integrable with zero integral. However, it is not difficult to show that f is not Riemann integrable (cf. Exercise 3.4).

The second corollary is another application of Theorem 3.2.

Corollary 3.2. *Let f be a bounded nonnegative measurable function on E. If $\int_E f = 0$, then $f(x) = 0$ a.e. on E.*

Proof. First we observe that

$$\{x \in E : f(x) > 0\} = \bigcup_{k=1}^{\infty} \{x \in E : f(x) > 1/k\}.$$

Suppose that $f(x) \neq 0$ on a set of positive measure. Then, because f is a nonnegative function, we must have $m(\{x \in E : f(x) > 0\}) > 0$. Therefore, there is n such that

$$\sigma = m(\{x \in E : f(x) > 1/n\}) > 0.$$

Let $A = \{x \in E : f(x) > 1/n\}$, $B = E \setminus A$. By Theorem 3.2,

$$\int_A f \geq \sigma/n \quad \text{and} \quad \int_B f \geq 0.$$

By Theorem 3.3,

$$\int_E f = \int_A f + \int_B f \geq \sigma/n,$$

contradicting our assumption that $\int_E f = 0$. $\qquad\square$

Now we turn our attention to the linearity properties of the Lebesgue integral.

Theorem 3.4. *Let f and g be bounded measurable functions with a common domain E. Then*

$$\int_E (f + g) = \int_E f + \int_E g.$$

Proof. Let (A, B) and (C, D) be open intervals such that

$$A < f(x) < B, \quad C < g(x) < D,$$

and let

$$A = y_0 < y_1 < \cdots < y_n = B \quad \text{and} \quad C = z_0 < z_1 < \cdots < z_m = D$$

be points defining partitions P and Q of intervals (A, B) and (C, D), respectively.

Let us consider sets

$$E_i(f, P) = \{x \in E : y_i \leq f(x) < y_{i+1}\}, \qquad 0 \leq i < n,$$
$$E_k(g, Q) = \{x \in E : z_k \leq g(x) < z_{k+1}\}, \qquad 0 \leq k < m,$$

and

$$T_{ik} = E_i(f, P) \cap E_k(g, Q), \quad 0 \leq i < n, \, 0 \leq k < m.$$

It is easy to see that the sets T_{ik} are pairwise disjoint and

$$E_i(f, P) = \bigcup_{k=1}^{m-1} T_{ik}, \quad E_k(g, Q) = \bigcup_{i=1}^{n-1} T_{ik}.$$

We have

$$y_i + z_k \leq f(x) + g(x) < y_{i+1} + z_{k+1}, \quad \text{for } x \in T_{ik}.$$

By Theorem 3.2,

$$(y_i + z_k)m(T_{ik}) \leq \int_{T_{ik}} (f + g) \leq (y_{i+1} + z_{k+1})m(T_{ik}).$$

If we add all the above inequalities, we obtain, by Theorem 3.3,

$$\sum_{i,k}(y_i + z_k)m(T_{ik}) \leq \int_E (f + g) \leq \sum_{i,k}(y_{i+1} + z_{k+1})m(T_{ik}).$$

We have

$$\sum_{i,k} y_i m(T_{ik}) = \sum_i y_i \sum_k m(T_{ik}) = \sum_i y_i m(E_i(f, P)) = s(f, P),$$
$$\sum_{i,k} z_k m(T_{ik}) = \sum_k z_k \sum_i m(T_{ik}) = \sum_k z_k m(E_k(g, Q)) = s(g, Q).$$

Similarly,

$$\sum_{i,k} y_{i+1} m(T_{ik}) = S(f, P) \quad \text{and} \quad \sum_{i,k} z_{k+1} m(T_{ik}) = S(g, Q).$$

Therefore,

$$s(f,P) + s(g,Q) \leq \int_E (f+g) \leq S(f,P) + S(g,Q).$$

By (3.1), we have

$$\int_E f + \int_E g \leq \int_E (f+g),$$

and by (3.2),

$$\int_E (f+g) \leq \int_E f + \int_E g.$$

The desired result follows from the last two displayed inequalities.   □

Theorem 3.5. *Let f be a bounded measurable function on E. Then*

$$\int_E (cf) = c \int_E f,$$

for an arbitrary constant c.

Proof. The assertion is trivial for $c = 0$. Suppose that $c > 0$ and let $y_0 < y_1 < \cdots < y_n$ be points defining a partition P of an interval containing the range of f. As before, let

$$E_k(f,P) = \{x \in E : y_k \leq f(x) < y_{k+1}\}, \quad 0 \leq k < n.$$

Clearly,

$$c\,y_k \leq cf(x) < c\,y_{k+1}, \quad \text{for } x \in E_k(f,P).$$

By Theorem 3.2,

$$c\,y_k m(E_k(f,P)) \leq \int_{E_k} (cf) \leq c\,y_{k+1} m(E_k(f,P)),$$

and by Theorem 3.3

$$c \cdot s(f,P) \leq \int_E (cf) \leq c \cdot S(f,P).$$

By (3.1) and (3.2), we have

$$c \int_E f \leq \int_E (cf) \leq c \int_E f,$$

that is, $\int_E (cf) = c \int_E f$.

For $c < 0$ we have, by Theorem 3.4 and the previous case,

$$0 = \int_E (cf + (-c)f) = \int_E (cf) + (-c)\int_E f,$$

and the result follows.   □

The results of the last two theorems often combined into the linearity property of integration:

Theorem 3.6. *Let f and g be bounded measurable functions on E and α and β be arbitrary constants. Then*

$$\int_E (\alpha f + \beta g) = \alpha \int_E f + \beta \int_E g.$$

Another important property of the Lebesgue integral is its monotonicity.

Theorem 3.7. *If f and g are bounded measurable functions on E and*

$$f(x) \leq g(x), \quad x \in E,$$

then

$$\int_E f \leq \int_E g.$$

Proof. It suffices to note that $g(x) - f(x) \geq 0$, for all $x \in E$, and apply Theorem 3.6 and the result of Exercise 3.1b.   □

The last theorem of this section is an analog of the triangle inequality for real numbers, $|\alpha + \beta| \leq |\alpha| + |\beta|$.

Theorem 3.8. *If f is a bounded measurable function on E, then*

$$\left| \int_E f \right| \leq \int_E |f|.$$

Proof. By Corollary 2.1, $|f|$ is a measurable function. Because

$$-|f(x)| \leq f(x) \leq |f(x)|, \quad \text{for } x \in E,$$

we have, by the linearity and monotonicity of the Lebesgue integral,

$$-\int_E |f| \leq \int_E f \leq \int_E |f|,$$

and the result follows.   □

3.3 Convergence

Let (f_n) be a uniformly convergent sequence of bounded measurable functions with the same domain E and let $f = \lim f_n$. By the definition of uniform convergence (cf. Definition 2.7(iii)), for $\varepsilon = 1$ there is N such that

$$|f - f_n| < 1, \quad \text{for } n \geq N.$$

Therefore,

$$f_N - 1 < f < f_N + 1,$$

that is, f is a bounded function. By Theorem 2.32, f is a measurable function. The next theorem establishes the "passage of the limit under the integral sign" for the uniform convergence.

Theorem 3.9. *If (f_n) is a sequence of bounded measurable functions on E that converges uniformly with*

$$\lim f_n = f,$$

then

$$\lim \int_E f_n = \int_E f.$$

Proof. The assertion is trivial if $m(E) = 0$ (cf. Exercise 3.1c), so we assume that $m(E) > 0$. Let $\varepsilon > 0$ and choose N for which

$$|f - f_n| < \varepsilon/m(E), \quad \text{for all } n \geq N.$$

By the results from Sect. 3.2,

$$\left| \int_E f - \int_E f_n \right| = \left| \int_E (f - f_n) \right| \leq \int_E |f - f_n| \leq \frac{\varepsilon}{m(E)} \cdot m(E) = \varepsilon.$$

Therefore, $\lim \int_E f_n = \int_E f$.   □

The claim of Theorem 3.9 does not hold in general for sequences that converge only pointwise as the following example demonstrates.

Example 3.2. Let (f_n), $n > 2$, be a sequence of functions on $[0, 1]$ defined by

$$f_n(x) = \begin{cases} n^2 x, & \text{if } 0 \leq x \leq 1/n, \\ -n^2 x + 2n, & \text{if } 1/n < x \leq 2/n, \\ 0, & \text{if } 2/n < x \leq 1 \end{cases}$$

(see Fig. 3.1).

It is not difficult to verify that $\lim f_n(x) = 0$ for all $x \in [0, 1]$ and that $\int_0^1 f_n = 1$ (cf. Exercise 3.10). Thus, (f_n) converges to the zero function on $[0, 1]$, whereas $\lim \int_0^1 f_n = 1$. Hence,

$$1 = \lim \int_0^1 f_n \neq \int_0^1 \lim f_n = 0.$$

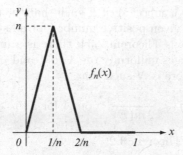

Figure 3.1. Sequence (f_n)

The result of the next theorem shows that the "passage of the limit under the integral sign" can be saved by assuming the uniform boundedness of the sequence under consideration.

Theorem 3.10. (The Bounded Convergence Theorem) *Let (f_n) be a sequence of measurable functions on E. Suppose that (f_n) is uniformly bounded on E, that is, there is $M \geq 0$ such that*

$$|f_n| \leq M, \quad \text{for all } n.$$

If (f_n) converges pointwise to f, then

$$\lim \int_E f_n = \int_E f.$$

Proof. As in the proof of Theorem 3.9, we may assume that E is a set of positive measure.

By Theorem 2.32, f is a measurable function. It is clear that $|f| \leq M$. Let A be a measurable subset of E and $n \in \mathbb{N}$. Then

$$\int_E f_n - \int_E f = \int_E (f_n - f) = \int_A (f_n - f) + \int_{E \setminus A} f_n + \int_{E \setminus A} (-f).$$

Observe that, by Theorems 3.8 and 3.2,

$$\left| \int_{E \setminus A} f_n \right| \leq \int_{E \setminus A} |f_n| \leq M \cdot m(E \setminus A)$$

and

$$\left| \int_{E \setminus A} (-f) \right| \leq \int_{E \setminus A} |f| \leq M \cdot m(E \setminus A).$$

Therefore,

$$\left| \int_E f_n - \int_E f \right| \leq \left| \int_A (f_n - f) \right| + \left| \int_{E \setminus A} f_n \right| + \left| \int_{E \setminus A} (-f) \right| \tag{3.3}$$

$$\leq \int_A |f_n - f| + 2M \cdot m(E \setminus A).$$

Our goal is to choose a subset A of E such that the right side of the above inequality is less than a given positive number for all sufficiently large n.

Let $\varepsilon > 0$. By Egorov's Theorem 2.34, there is a measurable subset A of E such that (f_n) converges uniformly on A to f and $m(E \setminus A) < \varepsilon/4M$. By uniform convergence, there is N such that

$$|f_n - f| < \frac{\varepsilon}{2\,m(F)}, \quad \text{on } A \text{ for all } n \geq N.$$

Therefore, by (3.3) and Theorem 3.2,

$$\left| \int_E f_n - \int_E f \right| < \frac{\varepsilon}{2\,m(E)} m(A) + 2M \frac{\varepsilon}{4M} \leq \varepsilon,$$

for all $n \geq N$. It follows that $\lim \int_E f_n = \int_E f$.   □

The following example shows that the assertion of the bounded convergence theorem does not hold for the Riemann integral.

Example 3.3. For $x \in [0,1]$, let us define

$$f_n(x) = \begin{cases} 1, & \text{if } x = p/q, \\ 0, & \text{otherwise}, \end{cases} \quad \text{where } q \leq n,\, p, q, n \in \mathbb{N}.$$

Each f_n is discontinuous on a finite set of points, so each f_n is Riemann integrable. However, the sequence (f_n) converges pointwise to Dirichlet's function which is not Riemann integrable (cf. Exercise 3.4).

3.4 Integration of Nonnegative Measurable Functions

As the first step towards extension of the Lebesgue integral to unbounded measurable functions, we consider nonnegative measurable functions on a measurable set.

For a given nonnegative measurable function f on E, we denote by H^f the set of bounded measurable functions h on E satisfying inequalities

$$0 \leq h \leq f \quad \text{on } E.$$

Definition 3.2. *Let f be a nonnegative measurable function on E. We define*

$$\int_E f = \sup \left\{ \int_E h : h \in H^f \right\}$$

and say that f is integrable *if $\int_E f < \infty$.*

We begin by establishing the linearity and monotonicity properties of the integral (cf. Theorems 3.6 and 3.7).

Theorem 3.11. *If f is a nonnegative measurable function on E and c is a positive constant, then*

$$\int_E (cf) = c \int_E f.$$

Proof. It is clear that $h \in H^{cf}$ if and only if $h/c \in H^f$. Therefore

$$\int_E (cf) = \sup \left\{ \int_E h : h \in H^{cf} \right\} = \sup \left\{ \int_E c \cdot (h/c) : h/c \in H^f \right\}$$

$$= c \cdot \sup \left\{ \int_E h/c : h/c \in H^f \right\} = c \int_E f,$$

because $c > 0$. $\qquad\qquad\qquad\qquad\qquad\qquad\qquad\qquad\qquad\qquad\qquad\qquad\Box$

Note that the result of this theorem does not hold in general if $c = 0$. Indeed, if $\int_E f = \infty$, then the left integral is zero, whereas the right side is indeterminate expression $0 \cdot \infty$. However, the result holds for an integrable f (that is, $\int_E f < \infty$), if $c = 0$.

Theorem 3.12. *Let f and g be nonnegative measurable functions on E. Then*

$$\int_E (f + g) = \int_E f + \int_E g.$$

Proof. For $h \in H^f$ and $k \in H^g$, we have $h + k \in H^{f+g}$. Therefore, by Theorem 3.6,

$$\int_E h + \int_E k = \int_E (h + k) \leq \int_E (f + g).$$

By taking the supremum over $h \in H^f$, $k \in H^g$ on the left side, we obtain

$$\int_E f + \int_E g \leq \int_E (f + g).$$

To prove the opposite inequality, we first observe that for any function $\ell \in H^{f+g}$, there are functions $h \in H^f$ and $k \in H^g$ such that

$$\ell = h + k.$$

Indeed let $h = \min\{f, \ell\}$. This function is measurable and bounded. Clearly, $0 \leq h \leq f$, so $h \in H^f$. Consider the function

$$k = \ell - h = \ell - \min\{f, \ell\} = \max\{\ell - f, 0\}.$$

Because $\ell \leq f + g$, we have $k \leq g$. Therefore, $k \in H^g$. We have

$$\int_E \ell = \int_E h + \int_E k \leq \int_E f + \int_E g.$$

By taking the supremum on the left side over all $\ell \in H^{f+g}$, we obtain the desired inequality

$$\int_E (f+g) \le \int_E f + \int_E g.$$

The result follows. □

In summary we have the following linearity property of integration for nonnegative measurable functions.

Theorem 3.13. *Let f and g be nonnegative measurable functions on E and α and β be arbitrary positive constants. Then*

$$\int_E (\alpha f + \beta g) = \alpha \int_E f + \beta \int_E g.$$

The next theorem establishes monotonicity of the integral.

Theorem 3.14. *If f and g are nonnegative measurable functions on E such that $f \le g$ on E. Then*

$$\int_E f \le \int_E g.$$

Proof. $H^f \subseteq H^g$, inasmuch as $f \le g$. Therefore, for any $h \in H^f$,

$$\int_E h \le \sup\left\{ \int_E k : k \in H^g \right\} = \int_E g.$$

By taking the supremum on the left side over all $h \in H^f$, we obtain the desired inequality. □

The result of Lemma 3.3 can be extended as the additivity property over domains of integration as follows:

Theorem 3.15. *Let f be a nonnegative integrable function on E. If*

$$E = \bigcup_{k=1}^{n} A_k,$$

where the sets A_k are measurable and pairwise disjoint, then

$$\int_E f = \sum_{k=1}^{n} \int_{A_k} f.$$

The proof is straightforward and left to the reader (cf. Exercise 3.13 and Theorem 3.18).

We establish now two important convergence properties of the integral of a nonnegative measurable function.

Theorem 3.16. (Fatou's Lemma) *Let* (f_n) *be a sequence of nonnegative measurable functions on* E. *If* (f_n) *converges pointwise to* f *a.e. on* E, *then*

$$\int_E f \leq \liminf \int_E f_n.$$

Proof. Let E_0 be the subset of E where (f_n) does not converge to f. Then $m(E_0) = 0$ and, by Theorem 3.15, $\int_E f_n = \int_{E \setminus E_0} f_n$ and $\int_E f = \int_{E \setminus E_0} f$. Note that (f_n) converges to f everywhere on $E \setminus E_0$. Thus, without loss of generality, we may assume in our proof that (f_n) converges everywhere to f.

First we note that, by Theorem 2.32, f is measurable and that $f \geq 0$. Let $h \in H^f$; that is, h is bounded, $h \leq M$ for some $M > 0$, and $0 \leq h \leq f$. Let us consider a sequence of functions defined by

$$h_n = \min\{h, f_n\} \quad \text{on } E.$$

The functions h_n are uniformly bounded by M,

$$h_n \leq M \quad \text{on } E, \text{ for all } n \in \mathbb{N},$$

and

$$\lim h_n = \min\{h, \lim f_n\} = \min\{h, f\} = h \quad \text{on } E.$$

By the bounded convergence theorem (Theorem 3.10),

$$\lim \int_E h_n = \int_E h.$$

We have $h_n \in H^{f_n}$, because $h_n \leq f_n$. Therefore, by the definition of $\int_E f_n$,

$$\int_E h_n \leq \int_E f_n.$$

Hence,

$$\int_E h = \lim \int_E h_n = \liminf \int_E h_n \leq \liminf \int_E f_n$$

(cf. Exercise 3.12). By taking the supremum on the left side over all $h \in H^f$, we obtain $\int_E f \leq \liminf \int_E f_n$. $\qquad \square$

The inequality in Fatou's Lemma may be strict as illustrated by Example 3.2. However, we have the equality in Fatou's Lemma if the sequence (f_n) is increasing.

Theorem 3.17. (The Monotone Convergence Theorem) *Let* (f_n) *be an increasing sequence of nonnegative measurable functions on* E. *If* (f_n) *converges pointwise to* f *a.e. on* E, *then*

$$\lim \int_E f_n = \int_E f.$$

Proof. As in the proof of Fatou's Lemma, we may assume that the convergence is on all of E.

Because $f_n \leq f$ on E, we have, by the monotonicity of integration,

$$\int_E f_n \leq \int_E f.$$

Therefore,

$$\limsup \int_E f_n \leq \int_E f.$$

On the other hand, by Fatou's Lemma,

$$\int_E f \leq \liminf \int_E f_n.$$

From the last two inequalities,

$$\limsup \int_E f_n \leq \int_E f \leq \liminf \int_E f_n.$$

The result follows from Exercise 1.31 and Theorem 1.5. $\square$

Corollary 3.3. *Let (u_n) be a sequence of nonnegative measurable functions on E. If*

$$f = \sum_{n=1}^{\infty} u_n,$$

a.e. on E, then

$$\int_E f = \sum_{n=1}^{\infty} \int_E u_n.$$

Proof. It suffices to apply the monotone convergence theorem to the sequence $f_n = \sum_{k=1}^{n} u_k$. $\square$

As an application of Corollary 3.3, we establish the *countable additivity* of the integral (cf. Theorem 3.3).

Theorem 3.18. *Let f be a nonnegative measurable function on E. If*

$$E = \bigcup_{i \in J} A_i,$$

where $\{A_i\}_{i \in J}$ is a finite or countable family of pairwise disjoint measurable sets, then

$$\int_E f = \sum_{i \in J} \int_{A_i} f.$$

Proof. Let us define

$$u_i(x) = \begin{cases} f(x), & \text{if } x \in A_i, \\ 0, & \text{if } x \in E \setminus A_i. \end{cases}$$

Then we have

$$f = \sum_{i \in J} u_i \quad \text{on } E.$$

By Corollary 3.3,

$$\int_E f = \sum_{i \in J} \int_E u_i.$$

It remains to be shown that

$$\int_E u_i = \int_{A_i} f, \quad \text{for all } i \in J.$$

For a given $i \in J$, let h_i be a bounded measurable function such that

$$0 \le h_i \le u_i \quad \text{on } E.$$

It is clear that $h_i = 0$ on $E \setminus A_i$. By Lemma 3.3,

$$\int_E h_i = \int_{A_i} h_i + \int_{E \setminus A_i} h_i = \int_{A_i} h_i.$$

The supremum of the integral on the left side of the above identity is $\int_E u_i$, whereas the supremum of the right side integral is $\int_{A_i} f$, because h_i is an arbitrary bounded measurable function such that

$$0 \le h_i \le f \quad \text{on } A_i.$$

Therefore, $\int_E u_i = \int_{A_i} f$, and the result follows. □

The last theorem of this section establishes an important "absolute continuity" of the Lebesgue integral (cf. Exercise 3.35).

Theorem 3.19. *Let f be a nonnegative function which is integrable over a set E. Then given $\varepsilon > 0$ there is $\delta > 0$ such that for every measurable subset A of E with $m(A) < \delta$ we have*

$$\int_A f < \varepsilon.$$

Proof. Let $\varepsilon > 0$. We define a sequence of functions on E by

$$f_n(x) = \min\{f(x), n\}, \qquad \text{for } n \in \mathbb{N}.$$

Then (f_n) is an increasing sequence of nonnegative measurable functions that converges pointwise to f. By the monotone convergence theorem (Theorem 3.17), the sequence $(\int_E f_n)$ converges to $\int_E f$. Therefore, there is N such that

$$\int_E f_N > \int_E f - \frac{\varepsilon}{2}.$$

For $\delta = \varepsilon/(2N)$ and a measurable set $A \subseteq E$ with $m(A) < \delta$, we have

$$\int_A f = \int_A (f - f_N) + \int_A f_N \le \int_E (f - f_N) + N m(A)$$
$$< \frac{\varepsilon}{2} + N m(A) < \varepsilon$$

(cf. Exercise 3.21 and Theorem 3.2.)   □

3.5 General Lebesgue Integral

For an arbitrary real-valued function f on E, we define its positive and negative parts by

$$f^+(x) = \max\{f(x), 0\} \quad \text{for all } x \in E$$

and

$$f^-(x) = \max\{-f(x), 0\} \quad \text{for all } x \in E,$$

respectively. Then f^+ and f^- are nonnegative functions and

$$f = f^+ - f^-, \quad |f| = f^+ + f^- \quad \text{on } E$$

(cf. Fig. 3.2).

Observe that f is measurable if and only if both f^+ and f^- are measurable (cf. Exercise 3.23).

Definition 3.3. *A measurable function f on E is said to be* integrable *over E provided that the functions f^+ and f^- are integrable over E. When this is so, the integral of f over E is defined by*

$$\int_E f = \int_E f^+ - \int_E f^-.$$

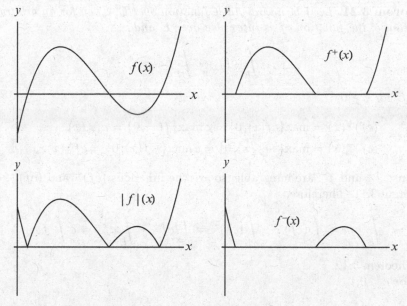

Figure 3.2. Graphs of functions f, f^+, $|f|$, and f^-

Theorem 3.20. *A measurable function f is integrable over E if and only if the function $|f|$ is integrable over E. In this case,*

$$\left| \int_E f \right| \le \int_E |f|.$$

(cf. Theorem 3.8)

Proof. (Necessity.) If f is integrable, then $|f| = f^+ + f^-$ is integrable, by the linearity property of integration for nonnegative functions.

(Sufficiency.) Suppose that $|f|$ is integrable over E. Because $0 \le f^+ \le |f|$ and $0 \le f^- \le |f|$, the functions f^+, f^- are integrable over E, by the monotonicity property of integration for nonnegative functions. Hence, f is integrable over E.

Finally,

$$\left| \int_E f \right| = \left| \int_E f^+ - \int_E f^- \right| \le \int_E f^+ + \int_E f^- = \int_E |f|,$$

by the triangle inequality for real numbers and the linearity of integration for nonnegative functions. $\qquad\square$

Now we extend the linearity and monotonicity properties of integration to arbitrary integrable functions.

Theorem 3.21. *Let f be an integrable function over E. Then for an arbitrary constant c, the function cf is integrable over E and*

$$\int_E (cf) = c \int_E f.$$

Proof. For $c \geq 0$,

$$(cf)^+(x) = \max\{cf(x), 0\} = c\max\{f(x), 0\} = cf^+(x),$$
$$(cf)^-(x) = \max\{-cf(x), 0\} = c\max\{-f(x), 0\} = cf^-(x).$$

Because f^+ and f^- are integrable, so are the functions $(cf)^+$ and $(cf)^-$, by Theorem 3.11. Therefore,

$$\int_E (cf) = \int_E (cf)^+ - \int_E (cf)^- = \int_E cf^+ - \int_E cf^- = c\int_E f,$$

by Theorem 3.12.

For $c < 0$,

$$(cf)^+(x) = \max\{cf(x), 0\} = -c\max\{-f(x), 0\} = -cf^-(x),$$
$$(cf)^-(x) = \max\{-cf(x), 0\} = -c\max\{f(x), 0\} = -cf^+(x),$$

and a similar argument shows that in this case again

$$\int_E (cf) = \int_E (cf)^+ - \int_E (cf)^- = \int_E (-c)f^- - \int_E (-c)f^+ = c\int_E f.$$

Hence, $\int_E (cf) = c\int_E f$ for any constant c. $\qquad\square$

Theorem 3.22. *Let f and g be integrable functions over E. Then the function $f + g$ is integrable over E and*

$$\int_E (f + g) = \int_E f + \int_E g.$$

Proof. By the linearity of integration for nonnegative functions and Theorem 3.20, $|f| + |g|$ is integrable over E. Because $|f + g| \leq |f| + |g|$, Theorem 3.14 tells us that the function $|f + g|$ is integrable and hence is $f + g$, by Theorem 3.20.

We have

$$(f + g)^+ - (f + g)^- = f + g = (f^+ - f^-) + (g^+ - g^-).$$

Therefore,

$$(f + g)^+ + f^- + g^- = (f + g)^- + f^+ + g^+.$$

By the linearity of integration for nonnegative functions, we obtain

$$\int_E (f+g)^+ + \int_E f^- + \int_E g^- = \int_E (f+g)^- + \int_E f^+ + \int_E g^+,$$

or, equivalently,

$$\int_E (f+g)^+ - \int_E (f+g)^- = \left(\int_E f^+ - \int_E f^- \right) + \left(\int_E g^+ - \int_E g^- \right).$$

Hence, $\int_E (f+g) = \int_E f + \int_E g$.   □

As in Sect. 3.2 (cf. Theorem 3.6) we formulate these results as the linearity property of integration.

Theorem 3.23. *Let f and g be integrable functions on E and α and β be arbitrary constants. Then $\alpha f + \beta g$ is integrable and*

$$\int_E (\alpha f + \beta g) = \alpha \int_E f + \beta \int_E g.$$

The next theorem establishes the monotonicity property of the general Lebesgue integral.

Theorem 3.24. *Let the functions f and g be integrable over E and*

$$f \le g \quad on\ E.$$

Then

$$\int_E f \le \int_E g.$$

Proof. Let us define $h = g - f$. Observe that h is a nonnegative measurable function on E which is integrable by Theorem 3.23. By the same theorem and monotonicity of integration for nonnegative functions,

$$\int_E g - \int_E f = \int_E (g-f) = \int_E h \ge 0.$$

The result follows.   □

We conclude this section by proving the following generalization of the bounded convergence theorem (Theorem 3.10).

Theorem 3.25. (The Dominated Convergence Theorem) *Let (f_n) be a sequence of measurable functions on E. Suppose that there is an integrable function g on E that dominates (f_n) on E in the sense that*

$$|f_n| \le g \quad on\ E\ for\ all\ n.$$

If (f_n) converges pointwise to f a.e. on E, then

$$\lim \int_E f_n = \int_E f.$$

Proof. As before, we may assume that the convergence takes place over all of E.

We have $|f| \leq g$, inasmuch as $f_n \to f$ pointwise on E and $|f_n| \leq g$ on E for all n. By the monotonicity of integration for nonnegative functions and Theorem 3.20, $|f| \leq g$ implies integrability of f.

Observe that $-g \leq f_n \leq g$ on E for all n, that is,

$$g - f_n \geq 0 \quad \text{and} \quad g + f_n \geq 0 \quad \text{on } E \text{ for all } n.$$

Because $g - f_n \geq 0$ on E for all n and $(g - f_n) \to (g - f)$ pointwise on E, we have, by the linearity property of integration and Fatou's Lemma,

$$\int_E g - \int_E f = \int_E (g - f) \leq \liminf \int_E (g - f_n) = \int_E g + \liminf \int_E (-f_n).$$

Hence,

$$\int_E f \geq \limsup \int_E f_n$$

(cf. Exercise 1.29a).

By the same reasoning, $g + f_n \geq 0$ and $(g + f_n) \to (g + f)$ imply

$$\int_E f \leq \liminf \int_E f_n.$$

By Exercise 1.29b,

$$\int_E f \leq \liminf \int_E f_n \leq \limsup \int f_n \leq \int_E f.$$

Now Theorem 1.5 tells us that $\lim \int_E f_n = \int_E f$. $\square$

3.6 Comparison of Riemann and Lebesgue Integrals

We begin by recalling some basic definitions and facts about the Riemann integral.

Let f be a bounded function on the closed interval $[a, b]$ and

$$P = \{x_0, x_1, \ldots, x_n\}, \qquad \text{where } x_0 = a \text{ and } x_n = b,$$

be a partition of this interval (cf. Sect. 3.1). We set $\triangle x_i = x_i - x_{i-1}$ for $i = 1, 2, \ldots, n$.

The *lower Riemann sum* of f over P is the number

$$L(f, P) = \sum_{i=1}^{n} m_i(f) \triangle x_i,$$

where

$$m_i = \inf\{f(x) : x \in [x_{i-1}, x_i]\}.$$

Similarly, the *upper Riemann sum* of f over P is the number

$$U(f, P) = \sum_{i=1}^{n} M_i(f) \bigtriangleup x_i,$$

where

$$M_i = \sup\{f(x) : x \in [x_{i-1}, x_i]\}.$$

Note that both numbers are well defined because f is a bounded function. The function f is said to be *Riemann integrable* on $[a, b]$ if

$$\lim_{\|P\| \to 0} L(f, P) = \lim_{\|P\| \to 0} U(f, P).$$

In this case, the common value of these two limits is the *Riemann integral* of f over $[a, b]$, $(R) \int_a^b f$.

For a given partition $P = \{x_0, x_1, \ldots, x_n\}$ of $[a, b]$, we define functions m and M on $[a, b]$ by

$$l_P(x) = m_i(f), \quad \text{if } x \in [x_{i-1}, x_i), 1 \leq i \leq n, \tag{3.4}$$

and

$$u_P(x) = M_i(f), \quad \text{if } x \in [x_{i-1}, x_i), 1 \leq i \leq n, \tag{3.5}$$

respectively, set $l_P(b) = u_P(b) = f(b)$, and observe that these functions are bounded and measurable, and their Lebesgue integrals are

$$(L) \int_a^b l_P = L(f, P) \quad \text{and} \quad (L) \int_a^b u_P = U(f, P). \tag{3.6}$$

In the rest of this section, we prove that a Riemann integrable function f on $[a, b]$ is Lebesgue integrable over the same interval and that its respective integrals are equal,

$$(R) \int_a^b f = (L) \int_a^b f.$$

Let f be a Riemann integrable function on $[a, b]$ and

$$P_1 \subseteq P_2 \subseteq \cdots \subseteq P_k \subseteq \cdots$$

be a nested family of partitions of the interval $[a, b]$ such that the sequence $(\|P_k\|)$ converges to zero. For a given $k \in \mathbb{N}$, we denote by l_k and u_k the functions l_P and u_P defined for the partition P_k by (3.4) and (3.5), respectively. Then we have

$$(L) \int_a^b l_k = L_k, \qquad (L) \int_a^b u_k = U_k,$$

where L_k and U_k are the lower and upper Riemann sums corresponding to P_k (cf. (3.6)). Inasmuch as the sequence (P_k) is nested, the sequence (l_k) is increasing and the sequence (u_k) is decreasing. Therefore each of these sequences converges pointwise to some bounded measurable functions, say l and u, respectively. We have

$$l \le f \le u, \tag{3.7}$$

because $l_k \le f \le u_k$. By the bounded convergence theorem (Theorem 3.10),

$$L_k \to (L) \int_a^b l \quad \text{and} \quad U_k \to (L) \int_a^b u. \tag{3.8}$$

Since f is Riemann integrable, both (L_k) and (U_k) converge to $(R) \int_a^b f$. It follows that

$$(R) \int_a^b f = (L) \int_a^b l = (L) \int_a^b u.$$

In particular, $(L) \int_a^b (u - l) = 0$. Because $u - l \ge 0$, Corollary 3.2 and (3.7) imply that $l = f = u$ a.e. on $[a, b]$. Therefore, f is measurable and, by (3.8),

$$(R) \int_a^b f = (L) \int_a^b f,$$

which completes the proof.

We proved that every Riemann integrable function is also Lebesgue integrable. However, the converse is not true. Indeed, Dirichlet's function (cf. Example 3.1) is Lebesgue integrable but not Riemann integrable (cf. Exercise 3.4). Thus the Lebesgue integral is a true generalization of the Riemann integral.

Notes

The notion of "partition" introduced in Sect. 3.1 differs from the one found in Sect. 1.1. It is always clear from the context which one is used.

The following passage is an excerpt from the talk "The development of the integral concept" given by Lebesgue at the Copenhagen Mathematical Society [Leb66, Part 2]. This informal motivation for the new concept of the integral appears in many later texts.

"One could say that, according to Riemann's procedure, one tried to add the indivisibles by taken them in the order in which they are furnished by the variation in x, like an unsystematic merchant who counts coins and bills at random in the order in which they come to hand, while we operate like a methodical merchant who says:

I have $m(E_1)$ pennies which are worth $1 \cdot m(E_1)$,
I have $m(E_2)$ nickels worth $5 \cdot m(E_2)$,
I have $m(E_3)$ dimes worth $10 \cdot m(E_3)$, etc.

Altogether then I have

$$S = 1 \cdot m(E_1) + 5 \cdot m(E_2) + 10 \cdot m(E_3) + \cdots .$$

The two procedures will certainly lead the merchant to the same result because no matter how much money he has there is only a finite number of coins or bills to count. But for us who must add an infinite number of indivisibles the difference between the two methods is of capital importance."

Exercises

3.1. Let f be a bounded measurable function on a set E. Show that

(a) If $f(x) = c$ for all $x \in E$, then

$$\int_E f = c \cdot m(E).$$

(b) If $f(x) \geq 0$ on E, then $\int_E f \geq 0$.
(c) If $m(E) = 0$, then $\int_E f = 0$.

3.2. Show that $\mu \leq \lambda \leq \nu$ implies

$$|\lambda| \leq \max\{|\mu|, |\nu|\}.$$

3.3. Let f and g be two bounded measurable functions on a set E. Prove that if $f(x) = g(x)$ a.e. on E, then

$$\int_E f = \int_E g.$$

Show that the converse statement does not hold.

3.4. Show that Dirichlet's function (Example 3.1) is not Riemann integrable.

3.5. Show that Dirichlet's function (Example 3.1) can be defined as the double limit

$$\lim_{m \to \infty} \lim_{n \to \infty} \cos^{2n}(m!\,\pi x).$$

3.6. Show that an increasing function on a closed interval is integrable and its integral is the same as the Riemann integral of the function.

3.7. Let A be a subset of a set E. The *characteristic function* of A, χ_A, is the function on E defined by

$$\chi_A(x) = \begin{cases} 1, & \text{if } x \in A, \\ 0, & \text{if } x \in E \setminus A. \end{cases}$$

Let f be a bounded measurable function on E. For a measurable subset A of E, show that $\int_A f = \int_E (f \cdot \chi_A)$.

3.8. Show that Theorem 3.9 follows from Theorem 3.10.

3.9. Let f be a bounded measurable function on $[a, b]$. Suppose that $\int_a^c f = 0$ for every $c \in [a, b]$. Prove that $f(x) = 0$ a.e. on $[a, b]$.

3.10. Show that a linear function on a closed interval is integrable and its integral is the same as the Riemann integral of the function. Establish the same result for a piecewise linear function on a closed interval.

3.11. Let f be a function on $[0, 1]$ defined by

$$f(x) = \begin{cases} x, & \text{if } x \in \mathbb{Q} \cap [0, 1], \\ -x, & \text{if } x \in [0, 1] \setminus \mathbb{Q}. \end{cases}$$

(Recall that $\mathbb{Q}$ is the set of rational numbers.) Is this function Riemann integrable? Lebesgue integrable?

3.12. Let (a_n) and (b_n) be two sequences of nonnegative numbers such that $a_n \leq b_n$ for all $n \in \mathbb{N}$. Prove that

$$\liminf a_n \leq \liminf b_n$$

(cf. Exercise 1.33).

3.13. Prove Theorem 3.15.

3.14. Show that the inequality in Fatou's Lemma is strict for the sequence

$$f_n(x) = (n + 1)x^n, \qquad x \in [0, 1]$$

(cf. Exercise 3.6).

3.15. Let (f_n) be a sequence of functions on $[0, 1]$ defined by

$$f_{2k+1}(x) = \begin{cases} 0, & \text{if } x \in [0, 1/3], \\ 1, & \text{otherwise,} \end{cases} \qquad f_{2k}(x) = \begin{cases} 1, & \text{if } x \in [0, 1/3], \\ 0, & \text{otherwise,} \end{cases}$$

$k = 0, 1, 2, \ldots$. Show that

$$\int_0^1 \liminf f_n < \liminf \int_0^1 f_n < \limsup \int_0^1 f_n < \int_0^1 \limsup f_n.$$

3.16. (Chebychesv's Inequality) Let f be a nonnegative measurable function on a set E. Prove that, for any $c > 0$,

$$m\{x \in E : f(x) \geq c\} \leq \frac{1}{c} \int_E f.$$

3.17. Let f be a nonnegative measurable function on E such that $\int_E f = 0$. Show that $f = 0$ a.e. on E.

3.18. Show that if f is integrable on E and if

$$E_n = \{x \in E : |f(x)| \geq n\},$$

then $\lim[n \cdot m(E_n)] = 0$ (cf. Exercise 3.16).

3.19. Let f be a nonnegative measurable function on E. Prove that f is integrable if and only if the family $\{m(E_n)\}_{n \in \mathbb{N}}$ is summable, where

$$E_n = \{x \in E : f(x) \geq n\}.$$

3.20. Let (f_n) be a sequence of nonnegative measurable functions on E that converges pointwise on E to an integrable function f. If $f_n \leq f$ on E for each n, show that

$$\lim \int_E f_n = \int_E f.$$

3.21. Let f be a nonnegative measurable function on E and let A be a measurable subset of E. Prove that

$$\int_A f \leq \int_E f.$$

3.22. Give an example of a measurable function f for which f^+ is integrable and f^- is not.

3.23. Prove that a function on a measurable set is measurable if and only if its positive and negative parts are both measurable.

3.24. Let f be integrable over E and A be a measurable subset of E. Show that

$$\int_A f = \int_E (f \cdot \chi_A)$$

(cf. Exercise 3.7).

3.25. Suppose that $\int_A f = 0$ for every measurable subset A of E. Show that $f(x) = 0$ a.e. on E.

3.26. Let f be a function defined by

$$f(x) = (-1)^n n, \qquad x \in E_n,$$

where

$$E_n = \left(\frac{1}{n+1}, \frac{1}{n}\right], \qquad n = 1, 2, \ldots.$$

Is this function integrable on $E = \bigcup_{n=1}^{\infty} E_n$?

3.27. Let f be a function defined by

$$f(x) = \sqrt{n}, \qquad x \in E_n,$$

where

$$E_n = \left(\frac{1}{n+1}, \frac{1}{n} \right], \qquad n = 1, 2, \ldots.$$

Is this function integrable on $E - \bigcup_{n=1}^{\infty} E_n$?

3.28. Let f be a function defined by

$$f(x) = (-1)^n, \qquad x \in E_n,$$

where

$$E_n = \left(\frac{1}{2^{n+1}}, \frac{1}{2^n} \right], \qquad n = 0, 1, 2, \ldots.$$

Evaluate $\int_E f$, where $E = \bigcup_{n=1}^{\infty} E_n$.

3.29. Let (f_n) be a sequence of integrable functions and let f be an integrable function such that

$$\lim \int_a^b |f_n(x) - f(x)| \, dx = 0.$$

Show that if $(f_n(x))$ converges a.e. on $[a, b]$, then $(f_n(x))$ converges to $f(x)$ a.e. on $[a, b]$.

3.30. Let $f_n(x) = nx^{n-1} - (n+1)x^n$, $0 < x < 1$. Show that

$$\int_0^1 \left(\sum_{n=1}^{\infty} f_n(x) \right) dx \neq \sum_{n=1}^{\infty} \left(\int_0^1 f_n(x) \, dx \right)$$

and

$$\sum_{n=1}^{\infty} \left(\int_0^1 |f_n(x)| \, dx \right) = \infty.$$

3.31. Let (f_n) be a sequence of measurable functions on E such that

$$\sum_{n=1}^{\infty} \left(\int_0^1 |f_n(x)| \, dx \right) < \infty.$$

Show that $\sum_{n=1}^{\infty} f_n$ is integrable and

$$\int_E \left(\sum_{n=1}^{\infty} f_n(x) \right) dx = \sum_{n=1}^{\infty} \left(\int_E f_n(x) \, dx \right).$$

3.32. Let f be a bounded measurable function. True or false: $|f|^p$ is integrable for any $p > 0$?

3.33. Let f be an integrable function. True or false: $|f|^p$ is integrable for any $p > 0$? (Cf. Exercise 3.32.)

3.34. Show that if f is integrable on E and

$$\left| \int_E f \right| = \int_E |f|,$$

then either $f \geq 0$ a.e. on E or $f \leq 0$ a.e. on E.

3.35. Prove that a measurable function on E is integrable over E if and only if for each $\varepsilon > 0$ there is $\delta > 0$ such that for every measurable subset A of E with $m(A) < \delta$ we have $\int_A |f| < \varepsilon$.

4

Differentiation and Integration

In the first three sections of this chapter, we investigate differentiability properties of functions of bounded variation. We begin by introducing the upper and lower derivatives of a function defined on a closed bounded interval $[a, b]$ in Sect. 4.1, where important properties of these derivatives are established (cf. Lemma 4.1). Functions of bounded variation are introduced and their properties are investigated in Sect. 4.2. The main result (Theorem 4.8) of Sect. 4.3 asserts that a function of a bounded variation on $[a, b]$ is differentiable almost everywhere on this interval.

In the remaining two sections, we present Lebesgue's versions (Theorems 4.10 and 4.14) of the two classical fundamental theorems of calculus (FTC):

FTC 1. *If f is continuous on $[a, b]$ and $F(x) = \int_a^x f(t)\, dt$, then F is continuously differentiable on $[a, b]$ and*

$$\frac{d}{dx} \int_a^x f(t)\, dt = f(x)$$

for each $x \in [a, b]$.

FTC 2. *If f is differentiable on $[a, b]$ and f' is integrable on $[a, b]$, then*

$$\int_a^x f'(t)\, dt = f(x) - f(a)$$

for each $x \in [a, b]$.

We set $\int_b^a f = - \int_a^b f$ for an integrable function f over the interval $[a, b]$.

S. Ovchinnikov, *Measure, Integral, Derivative: A Course on Lebesgue's Theory*, Universitext, DOI 10.1007/978-1-4614-7196-7_4, © Springer Science+Business Media New York 2013

4.1 Upper and Lower Derivatives

Let x be an inner point of a closed interval $[a, b]$ and let ε be a positive number. The set

$$B'_\varepsilon = \{y \in [a, b] : y \in (x - \varepsilon, x) \cup (x, x + \varepsilon)\}$$

is called a *punctured ball* (of radius ε, centered at x) in $[a, b]$. For a function f on $[a, b] \setminus \{x\}$, we define functions

$$M_\varepsilon = \sup\{f(y) : y \in B'_\varepsilon\} \quad \text{and} \quad m_\varepsilon = \inf\{f(y) : y \in B'_\varepsilon\}.$$

It is clear that M_ε is an increasing function of ε, whereas m_ε is a decreasing function of ε. Recall (cf. Sect. 1.2) that we set $M_\varepsilon = \infty$ if f is not bounded above on B'_ε and $m_\varepsilon = -\infty$ if f is not bounded below on B'_ε.

The *upper limit* of the function f as y tends to x is defined by

$$\varlimsup_{y \to x} f(y) = \inf\{M_\varepsilon : \varepsilon > 0\} = \inf_{\varepsilon > 0} \sup_{0 < |y - x| < \varepsilon} f(y)$$

and its *lower limit* by

$$\varliminf_{y \to x} f(y) = \sup\{m_\varepsilon : \varepsilon > 0\} = \sup_{\varepsilon > 0} \inf_{0 < |y - x| < \varepsilon} f(y).$$

Inasmuch as $m_\varepsilon \leq M_\varepsilon$ for all $\varepsilon > 0$, we have

$$\varliminf_{y \to x} f(y) \leq \varlimsup_{y \to x} f(y).$$

In the following four examples, the interval is $[-1, 1]$ and its inner point is $0 \in (-1, 1)$.

Example 4.1. For

$$f(x) = \begin{cases} -1, & \text{if } x \in [-1, 0), \\ 1, & \text{if } x \in (0, 1], \end{cases}$$

we have $M_\varepsilon = 1$ and $m_\varepsilon = -1$ for every $\varepsilon > 0$. Hence,

$$\varliminf_{y \to 0} f(y) = -1, \quad \varlimsup_{y \to 0} f(y) = 1.$$

Example 4.2. Let $f(x) = 1/x$ on $[-1, 1] \setminus \{0\}$. Then $M_\varepsilon = \infty$ and $m_\varepsilon = -\infty$ for every $\varepsilon > 0$. Therefore,

$$\varliminf_{y \to 0} f(y) = -\infty, \quad \varlimsup_{y \to 0} f(y) = \infty.$$

Example 4.3. Let $f(x) = 1/|x|$ on $[-1, 1] \setminus \{0\}$. Then $M_\varepsilon = \infty$ and $m_\varepsilon = 1/\varepsilon$ for every $\varepsilon > 0$. Therefore,

$$\varliminf_{y \to 0} f(y) = \varlimsup_{y \to 0} f(y) = \infty.$$

Example 4.4. Let $f(x) = x$ on $[\ 1,1] \setminus \{0\}$. Then $M_\varepsilon - \varepsilon$ and $m_\varepsilon - -\varepsilon$ for every $\varepsilon > 0$. Therefore,

$$\lim_{y \to 0} f(y) = \overline{\lim_{y \to 0}} f(y) = 0.$$

Theorem 4.1. *A function f on $[a,b] \setminus \{x\}$ has a limit at x if and only if $\underline{\lim}_{y \to x} f(y)$ and $\overline{\lim}_{y \to x} f(y)$ are finite and equal. Then*

$$\lim_{y \to x} f(y) = \underline{\lim_{y \to x}} f(y) = \overline{\lim_{y \to x}} f(y).$$

Proof. (Necessity.) Suppose that $\lim_{y \to x} f(y) = L$, that is, for any given $\varepsilon > 0$ there is $\delta > 0$ such that

$$L - \varepsilon < f(y) < L + \varepsilon, \qquad \text{for all } y \in B'_\delta.$$

It follows that

$$L - \varepsilon < M_{\delta'} \leq L + \varepsilon,$$

for any $\delta' < \delta$. Inasmuch as M_δ is an increasing function,

$$L - \varepsilon < \overline{\lim_{y \to x}} f(y) = \inf_{0 < \delta' < \delta} M_{\delta'} \leq L + \varepsilon.$$

Because ε is an arbitrary positive number, $\overline{\lim}_{y \to x} f(y) = L$.

A similar argument shows that $\underline{\lim}_{y \to x} f(y) = L$.

(Sufficiency.) Suppose that the upper and lower limits are finite and equal. Then there is a real number L such that

$$\underline{\lim_{y \to x}} f(y) = \overline{\lim_{y \to x}} f(y) = L,$$

that is,

$$\inf\{M_\delta : \delta > 0\} = L \quad \text{and} \quad \sup\{m_\delta : \delta > 0\} = L.$$

Therefore, for a given $\varepsilon > 0$, there is $\delta' > 0$ such that

$$L \leq M_{\delta'} < L + \varepsilon$$

and $\delta'' > 0$ such that

$$L - \varepsilon < m_{\delta''} \leq L.$$

Let $\delta = \min\{\delta', \delta''\}$. Then, by the monotonicity of functions m_δ and M_δ,

$$L - \varepsilon < m_\delta \leq M_\delta < L + \varepsilon.$$

Hence,

$$L - \varepsilon < f(y) < L + \varepsilon, \qquad \text{for all } y \in B'_\delta,$$

that is, $\lim_{y \to x} f(y) = L$. □

Now, let f be a function defined on the entire interval $[a, b]$ and let x be a fixed inner point of $[a, b]$. Then the function

$$\frac{f(y) - f(x)}{y - x}$$

(as a function of y) is defined on $[a, b] \setminus \{x\}$. The (possibly extended) number

$$\overline{D}f(x) = \varlimsup_{y \to x} \frac{f(y) - f(x)}{y - x}$$

is called the *upper derivative* of f at x. Similarly, the *lower derivative* of f at x is the number

$$\underline{D}f(x) = \varliminf_{y \to x} \frac{f(y) - f(x)}{y - x}.$$

If the limit

$$\lim_{y \to x} \frac{f(y) - f(x)}{y - x}$$

exists, it is called the *derivative* of f at x and denoted by $f'(x)$ or $Df(x)$. In this case the function f is said to be *differentiable* at x. The following theorem follows immediately from Theorem 4.1.

Theorem 4.2. *A function f on $[a, b]$ is differentiable at $x \in (a, b)$ if and only if its lower and upper derivatives at x are equal finite numbers. Then*

$$Df(x) = \underline{D}f(x) = \overline{D}f(x).$$

Note that we always have $\underline{D}f(x) \leq \overline{D}f(x)$.

The results of the next lemma are weak forms of the mean value theorem for the derivative in analysis. They are essential in proving the main theorem of Sect. 4.3.

Lemma 4.1. *Let f be a function on $[a, b]$ which is continuous at a point $x \in (a, b)$, and let A be a positive constant, $A > 0$.*

(i) *If $\overline{D}f(x) > A$, then there are numbers a_x, b_x such that*

$$a < a_x < x < b_x < b$$

and

$$\frac{f(b_x) - f(a_x)}{b_x - a_x} > A.$$

(ii) *If $\underline{D}f(x) < -A$, then there are numbers a_x, b_x such that*

$$a < a_x < x < b_x < b$$

and

$$\frac{f(b_x) - f(a_x)}{b_x - a_x} < -A.$$

(iii) *Continuity of f at x is a necessary condition for assertions* (i) *and* (ii).

Proof.

(i) Because

$$\overline{D}f(x) = \inf_{\varepsilon>0} \sup_{0<|y-x|<\varepsilon} \frac{f(y) - f(x)}{y - x} > A,$$

we have

$$\sup_{0<|y-x|<\varepsilon} \frac{f(y) - f(x)}{y - x} > A, \qquad \text{for all } \varepsilon > 0.$$

Therefore, for $\varepsilon = \min\{x - a, b - x\}$, there is $y \neq x$ in (a, b) such that

$$\frac{f(y) - f(x)}{y - x} > A. \tag{4.1}$$

If $y < x$, we define $a_x = y$. Otherwise, we define $b_x = y$. Because the arguments in both cases are similar, we consider the latter case only. Inasmuch as f is continuous at x, we have, by (4.1),

$$\lim_{z \to x^-} \frac{f(b_x) - f(z)}{b_x - z} = \frac{f(b_x) - f(x)}{b_x - x} > A,$$

where $z < x < b_x$. Therefore, there is $z = a_x$ such that

$$\frac{f(b_x) - f(a_x)}{b_x - a_x} > A.$$

(ii) Because $\underline{D}f(x) < -A$ is equivalent to $\overline{D}[-f(x)] > A$ (cf. Exercise 4.2), we have, by part (i),

$$\frac{-f(b_x) + f(a_x)}{b_x - a_x} > A,$$

which is equivalent to

$$\frac{f(b_x) - f(a_x)}{b_x - a_x} < -A.$$

(iii) Let f be a function on $[-1, 1]$ defined by

$$f(x) = \begin{cases} x + 1, & \text{if } -1 \le x \le 0, \\ -1, & \text{if } 0 < x \le 1. \end{cases}$$

Observe that f is discontinuous at $x = 0$. Then

$$\frac{f(y) - f(0)}{y - 0} = \frac{f(y) - 1}{y} = \begin{cases} 1, & \text{if } -1 \le y < 0, \\ -1/y, & \text{if } 0 < y \le 1. \end{cases}$$

Hence,

$$\overline{D}f(0) = \inf_{\varepsilon>0} \sup_{0<|y|<\varepsilon} \frac{f(y) - f(0)}{y - 0} = 1.$$

On the other hand, for any $-1 \le z < 0$, $0 < y \le 1$, we have

$$\frac{f(y) - f(z)}{y - z} = \frac{-2 - z}{y - z} < 0.$$

Thus (i) does not hold for f if, say, $A = 1/2$. Similarly, (ii) does not hold for $-f$ and $A = -1/2$. $\qquad\square$

4.2 Functions of Bounded Variation

Let f be a function on $[a, b]$. Observe that over a subinterval $[c, d]$ of $[a, b]$ the values of the function f "change" or "vary" from $f(c)$ to $f(d)$, so the "change" in values is $|f(d) - f(c)|$. This observation motivates the following definition.

Definition 4.1. *Let f be a function on the interval $[a, b]$ and let $P = \{x_0, x_1, \ldots, x_n\}$ be a partition of this interval, that is,*

$$a = x_0 < x_1 < \cdots < x_n = b$$

(cf. Sect. 3.1). The variation of f *on $[a, b]$ with respect to P is given by*

$$V_a^b(P, f) = \sum_{k=1}^{n} |f(x_k) - f(x_{k-1})|.$$

If there is a constant C such that $V_a^b(P, f) < C$ for every partition P of $[a, b]$, then we say that f is a function of bounded variation, *a* BV-function *for short, on $[a, b]$ and write*

$$V_a^b(f) = \sup_{P} V_a^b(P, f).$$

In this case, the quantity $V_a^b(f)$ is called the total variation *of f over the interval $[a, b]$.*

Example 4.5. Consider the continuous function

$$f(x) = \begin{cases} x\cos\frac{1}{x}, & \text{if } x \neq 0, \\ 0, & \text{if } x = 0, \end{cases} \quad \text{on } [0, 1/\pi].$$

For the partition

$$P = \left\{ 0, \frac{1}{n\pi}, \frac{1}{(n-1)\pi}, \ldots, \frac{1}{2\pi}, \frac{1}{\pi} \right\}$$

we have

$$V_0^{1/\pi}(P, f) = \frac{1}{n\pi} + \sum_{k=1}^{n-1} \left(\frac{1}{(n-k+1)\pi} + \frac{1}{(n-k)\pi} \right)$$

$$= \frac{2}{\pi} \left(1 + \frac{1}{2} + \cdots + \frac{1}{n} \right) - \frac{1}{\pi}.$$

Inasmuch as the harmonic series diverges, the function f is not of a bounded variation.

Example 4.6. Let

$$f(x) = \begin{cases} x^2 \sin \frac{1}{x}, & \text{if } x \neq 0, \\ 0, & \text{if } x = 0, \end{cases} \quad \text{on } [-1, 1].$$

It is easy to verify that f is differentiable over the interval $[-1, 1]$ with $|f'(x)| \leq 3$ for $x \in (-1, 1)$. Therefore,

$$\sum_{k=1}^{n} |f(x_k) - f(x_{k-1})| \leq 3 \sum_{k=1}^{n} |x_k - x_{k-1}| = 3(b - a),$$

for any partition $P = \{x_0, x_1, \ldots, x_n\}$ of $[a, b]$. Thus f is a BV-function.

Example 4.7. Let f be an increasing function on $[a, b]$. Then

$$V_a^b(P, f) = \sum_{k=1}^{n} |f(x_k) - f(x_{k-1})| = f(b) - f(a),$$

for any partition of $[a, b]$. It follows that $V_a^b(f) = f(b) - f(a)$. If f is a decreasing function, then $V_a^b(f) = f(a) - f(b)$. Thus

$$V_a^b(f) = |f(b) - f(a)|$$

for any monotone function on $[a, b]$.

The triangle inequality implies that for an arbitrary function f on $[a, b]$,

$$\sum_{k=1}^{n} |f(x_k) - f(x_{k-1})| \geq |f(b) - f(a)|.$$

Thus $V_a^b(P, f) \geq |f(b) - f(a)|$ for any partition P of $[a, b]$. The following lemma gives a better estimate for the lower bound of a variation.

Lemma 4.2. *Let f be a function on $[a, b]$, $P = \{x_0, x_1, \ldots, x_n\}$ be a partition of $[a, b]$, S be a nonempty subset of $\{1, 2, \ldots, n\}$, and A be a positive number.*

(i) *If $f(a) \geq f(b)$ and*

$$\frac{f(x_k) - f(x_{k-1})}{x_k - x_{k-1}} > A, \qquad \text{for each } k \in S,$$

then

$$V_a^b(P, f) = \sum_{k=1}^{n} |f(x_k) - f(x_{k-1})| > |f(b) \quad f(a)| + A \cdot L,$$

where $L = \sum_{k \in S}(x_k - x_{k-1})$.
(ii) *The same inequality holds if $f(a) \leq f(b)$ and*

$$\frac{f(x_k) - f(x_{k-1})}{x_k - x_{k-1}} < -A, \qquad \text{for each } k \in S,$$

Proof.

(i) We have

$$|f(b) - f(a)| = f(a) - f(b) = \sum_{k=1}^{n}(f(x_{k-1}) - f(x_k))$$

$$= -\sum_{k \in S}(f(x_k) - f(x_{k-1})) + \sum_{k \notin S}(f(x_{k-1}) - f(x_k))$$

$$< -A \cdot L + \sum_{k \notin S}(f(x_{k-1}) - f(x_k))$$

$$\leq -A \cdot L + \sum_{k=1}^{n}|f(x_{k-1}) - f(x_k)| = -A \cdot L + V_a^b(P, f).$$

The desired inequality follows.
(ii) We obtain the result by applying the assertion in (i) to the function $-f$.

□

Note that if f is a BV-function on $[a, b]$, then it is a BV-function on any closed subinterval of $[a, b]$ (cf. Exercise 4.4).

Theorem 4.3. *Let f be a BV-function over $[a, b]$ and let c be an inner point of $[a, b]$. Then*

$$V_a^b(f) = V_a^c(f) + V_c^b(f).$$

Proof. Let P' be a partition of $[a, c]$ and P'' be a partition of $[c, b]$. Then $P' \cup P''$ is a partition of $[a, b]$ and

$$V_a^c(P', f) + V_c^b(P'', f) = V_a^b(P' \cup P'', f) \leq V_a^b(f).$$

By taking the supremum over P' and P'' separately, we obtain

$$V_a^c(f) + V_c^b(f) \leq V_a^b(f).$$

To obtain an opposite inequality, let P be an arbitrary partition of $[a,b]$. Then $c \in [x_{k-1}, x_k]$ for some $1 \leq k \leq n$. Because

$$|f(c) - f(x_{k-1})| + |f(x_k) - f(c)| \geq |f(x_k) - f(x_{k-1})|,$$

we have

$$V_a^b(P \cup \{c\}, f) \geq V_a^b(P, f).$$

Let P' and P'' be restrictions of $P \cup \{c\}$ to the intervals $[a, c]$ and $[c, b]$, respectively. Then

$$V_a^b(P, f) \leq V_a^b(P \cup \{c\}, f) = V_a^c(P', f) + V_c^b(P'', f)$$
$$\leq V_a^c(f) + V_c^b(f).$$

Therefore,

$$V_a^b(f) \leq V_a^c(f) + V_c^b(f)$$

and the result follows. $\qquad\qquad\square$

Corollary 4.1. $V_a^x(f)$ *is an increasing function on* $[a, b]$.

Proof. For $u < v$ in $[a, b]$ we have, by Theorem 4.3,

$$V_a^v(f) = V_a^u(f) + V_u^v(f).$$

Because $V_u^v(f) \geq 0$, we have $V_a^u(f) \leq V_a^v(f)$. $\qquad\qquad\square$

Theorem 4.4. (Jordan Decomposition Theorem) *A function f is a BV-function on $[a, b]$ if and only if it is the difference of two increasing functions on $[a, b]$.*

Proof. (Necessity.) Let us define $g(x) = V_a^x(f)$. By Corollary 4.1, $g(x)$ is an increasing function on $[a, b]$. Because $f = g - (g - f)$, it suffices to show that $h = g - f$ is an increasing function. For

$$a \leq u < v \leq b$$

we have

$$h(v) - h(u) = g(v) - g(u) - (f(v) - f(u))$$
$$= V_u^v(f) - (f(v) - f(u)) \geq 0,$$

because $V_u^v(f) \geq |f(v) - f(u)|$. (Recall that $V_u^v(P, f) \geq |f(v) - f(u)|$ for any partition P of $[u, v]$.) Hence h is an increasing function.

(Sufficiency.) Suppose that $f = g - h$ where g and h are increasing functions on $[a, b]$. Let $P = \{x_0, x_1, \ldots, x_n\}$ be a partition of $[a, b]$. By the triangle inequality and telescoping, we have

$$
\begin{aligned}
V_a^b(P, f) &= \sum_{k=1}^{n} |f(x_k) - f(x_{k-1})| \\
&\leq \sum_{k=1}^{n} |g(x_k) - g(x_{k-1})| + \sum_{k=1}^{n} |h(x_k) - h(x_{k-1})| \\
&= [g(b) - g(a)] + [h(b) - h(a)].
\end{aligned}
$$

Hence, f is a BV-function. $\qquad\square$

Corollary 4.2. *Let f be a BV-function on $[a, b]$. Then f has the following explicit expression as the difference of two increasing functions on $[a, b]$:*

$$ f = V_a^x(f) - [V_a^x(f) - f]. $$

Theorem 4.5. *A BV-function on $[a, b]$ is continuous except possibly at a finite or countable number of points in $[a, b]$.*

Proof. By the Jordan Decomposition Theorem, it suffices to prove the claim for an increasing function f on $[a, b]$. Because f is increasing, it has a limit from the left and from the right at any given point $x \in (a, b)$. We define

$$ f(x^-) = \lim_{y \to x^-} f(y) = \sup\{f(y) : a < y < x\} $$

and

$$ f(x^+) = \lim_{y \to x^+} f(y) = \inf\{f(y) : x < y < b\}. $$

Because f is increasing, $f(x^-) \leq f(x^+)$. The function f is discontinuous at x if and only if $f(x^-) < f(x^+)$, in which case we define the open "jump" interval

$$ J(x) = \{y : f(x^-) < y < f(x^+)\}. $$

The jump intervals are pairwise disjoint. Inasmuch as each jump interval contains a rational number, we can have no more than a countable set of these intervals. It follows that f cannot have more than a countable set of discontinuities. $\qquad\square$

4.3 Differentiability of BV-Functions

We begin by establishing some auxiliary results about open sets and families of intervals.

Lemma 4.3. (Lindelöf Property) *Let $[G_i]_{i \in J}$ be an infinite family of open sets. Then there is a countable subfamily $\{G_i\}_{i \in J_0}$ (J_0 is a countable subset of J) such that*

$$\bigcup_{i \in J} G_i = \bigcup_{i \in J_0} G_i.$$

Proof. Let $G = \bigcup_{i \in J} G_i$ and $x \in G$. Then $x \in G_i$ for some i in J. Inasmuch as G_i is open, there is an open interval I_x such that $x \in I_x \subseteq G_i$. We may assume that the endpoints of I_x are rational numbers (cf. Exercise 4.18). Thus we obtained a countable family of open intervals with the union G. For each interval I_x, we select an open set in $\{G_i\}_{i \in J}$ that contains it. The result follows. □

Lemma 4.4. *Let E be a subset of $[a, b]$ and $\{I_i\}_{i \in J}$ be a family of subintervals of $[a, b]$ such that*

$$E \subseteq \bigcup_{i \in J} I_i.$$

Then there is a nonempty finite subfamily $\{I_i\}_{i \in J_0}$ of $\{I_i\}_{i \in J}$ such that

$$\frac{1}{2} m^*(E) \leq \sum_{i \in J_0} m(I_i).$$

Proof. The assertion is trivial if $m^*(E) = 0$, so we assume that $m^*(E) > 0$.

By Lemma 4.3, we may assume that the set J is at most countable. By Theorems 2.12 and 2.13,

$$m^*(E) \leq m^* \left(\bigcup_{i \in J} I_i \right) \leq \sum_{i \in J} m(I_i).$$

If $\{m(I_i)\}_{i \in J}$ is not summable, then there is a finite subset $J_0 \subseteq J$ such that

$$\sum_{i \in J_0} m(I_i) > m^*(E)$$

and the result follows immediately. Otherwise, for

$$\varepsilon = \frac{1}{2} m^*(E) > 0$$

there is a finite set $J_0 \subseteq J$ such that

$$\left| \sum_{i \in J} m(I_i) - \sum_{i \in J_0} m(I_i) \right| < c - \frac{1}{2} m^*(E).$$

Therefore,

$$\sum_{i \in J_0} m(I_i) > \sum_{i \in J} m(I_i) - \frac{1}{2}m^*(E)$$

$$\geq m^*(E) - \frac{1}{2}m^*(E) = \frac{1}{2}m^*(E),$$

which is the desired result.   □

Note that we can replace $1/2$ in Lemma 4.4 by any number $0 < \lambda < 1$. However, we cannot choose $\lambda = 1$ as the following example illustrates.

Example 4.8. Let $E = (0, 1]$ be a subset of $[0, 1]$ and let

$$E = \bigcup_{i \in \mathbb{N}} \left(\frac{1}{i+1}, \frac{1}{i} \right].$$

Clearly, for any finite subset J_0 of $\mathbb{N}$, we have

$$m^*(E) > \sum_{i \in J_0} m\left(\frac{1}{i+1}, \frac{1}{i} \right].$$

Lemma 4.5. *Let $\{I_1, \ldots, I_n\}$ be a set of bounded intervals. There is a nonempty subset $S \subseteq \{1, \ldots, n\}$ such that the intervals I_i with $i \in S$ are pairwise disjoint and*

$$\sum_{i \in S} m(I_i) \geq \frac{1}{3}m\left(\bigcup_{i=1}^{n} I_i \right).$$

Proof. The proof is by induction on n. The claim is trivial for $n = 1$.

Let $n > 1$ and suppose that the assertion holds for all integers less than n. We may assume that I_1 is the interval of maximum length. Let

$$T = \{i \in \{1, \ldots, n\} : I_i \cap I_1 \neq \varnothing\}$$

and $T' = \{1, \ldots, n\} \setminus T$. Note that $1 \in T$, so the set T is not empty.

If $T = \{1, \ldots, n\}$, then $\bigcup_{i=1}^{n} I_i$ is an interval containing I_1 and

$$m(I_1) \geq \frac{1}{3}m\left(\bigcup_{i=1}^{n} I_i \right)$$

inasmuch as I_1 is of maximum length (cf. Exercise 4.19). Thus the claim holds for $S = \{1\}$.

If $T \neq \{1, \ldots, n\}$, then $\bigcup_{i \in T} I_i$ is an interval containing I_1 and

$$m(I_1) \geq \frac{1}{3}m\left(\bigcup_{i \in T} I_i \right).$$

(cf. Exercise 4.19).

The set $\{I_i : i \in T'\}$ is not empty and consists of intervals that are disjoint from I_1. By the induction hypothesis, there is a nonempty subset S' of T' such that the intervals in $\{I_i : i \in S'\}$ are pairwise disjoint and

$$\sum_{i \in S'} m(I_i) \geq \frac{1}{3} m\left(\bigcup_{i \in T'} I_i \right).$$

Let $S = S' \cup \{1\}$. By adding the last two displayed inequalities we obtain

$$\sum_{i \in S} m(I_i) \geq \frac{1}{3} \left[m\left(\bigcup_{i \in T} I_i \right) + m\left(\bigcup_{i \in T'} I_i \right) \right] \geq \frac{1}{3} m\left(\bigcup_{i=1}^{n} I_i \right).$$

The result follows by induction.   □

Lemma 4.6. *Let a bounded set E be the union of an at most countable family of sets $\{E_i\}_{i \in J}$,*

$$E = \bigcup_{i \in J} E_i.$$

Then E is a set of measure zero if and only if each set E_i is of measure zero.

Proof. (Necessity.) Suppose that $m(E) = 0$ and there is $i \in J$ such that E_i is not of measure zero. Then $m^*(E_i) > 0$ (cf. Exercise 4.20). Because $E_i \subseteq E$, we have

$$0 < m^*(E_i) \leq m^*(E),$$

contradicting our assumption that E is a set of measure zero (cf. Exercise 4.20).

(Sufficiency.) If $m(E_i) = 0$ for all $i \in J$, then, by Theorem 2.13,

$$m^*(E) \leq \sum_{i \in J} m^*(E_i) = 0.$$

Inasmuch as $0 \leq m_*(E) \leq m^*(E) \leq 0$, we have $m(E) = 0$.   □

Theorem 4.6. *Let f be a BV-function on $[a, b]$ and*

$$E = \{x \in (a, b) : \overline{D}f(x) \neq \underline{D}f(x)\}.$$

Then E is a set of measure zero.

Proof. Because $\overline{D}f(x) \geq \underline{D}f(x)$ on (a, b), we have

$$E = \{x \in (a, b) : \overline{D}f(x) > \underline{D}f(x)\}.$$

Let

$$E_c = \{x \in (a, b) : f \text{ is continuous at } x\}.$$

By Theorem 4.5, $(a, b) \setminus E_c$ is at most countable. Therefore it suffices to show that the set

$$E' = \{x \in E_c : \overline{D}f(x) > \underline{D}f(x)\}$$

is of measure zero (cf. Exercise 4.21).

For rational numbers $A > 0$ and B, we define

$$E_{A,B} = \{x \in E_c : \overline{D}f(x) > B + A > B - A > \underline{D}f(x)\}.$$

Then

$$E' = \bigcup_{A,B} E_{A,B}$$

(cf. proof of Theorem 2.30). By Lemma 4.6, to show that E' is a set of measure zero it suffices to show that all sets $E_{A,B}$ are of measure zero. We do it by contradiction.

Suppose that there are two rational numbers $A > 0$ and B such that $E_{A,B}$ is not a set of measure zero and let $G = E_{A,B}$. Thus we assume that $m^*(G) > 0$ (cf. Exercise 4.20). In what follows, we show that this condition contradicts our assumption that f is a BV-function on $[a, b]$.

Because f is a BV-function, the function $g(x) = f(x) - Bx$ is also a BV-function (cf. Exercise 4.6). Note that the set E_c is the set of points in (a, b) where the function g is continuous. We have

$$\overline{D}g(x) = \overline{D}f(x) - B \quad \text{and} \quad \underline{D}g(x) = \underline{D}f(x) - B.$$

Therefore,

$$G = \{x \in E_c : \overline{D}g(x) > A > -A > \underline{D}g(x)\}.$$

Let $P = \{x_0, x_1, \ldots, x_n\}$, where

$$a = x_0 < x_1 < \cdots < x_n = b,$$

be a partition of $[a, b]$ such that

$$V_a^b(P, g) = \sum_{i=1}^n |g(x_i) - g(x_{i-1})| > V_a^b(g) - A \cdot \frac{1}{6} m^*(G). \tag{4.2}$$

By the definition of the total variation $V_a^b(g)$, such a partition exists, because $A > 0$ and $m^*(G) > 0$. For a given $x \in G \setminus P$, we have $x \in (x_{i-1}, x_i)$ for some $1 \leq i \leq n$. Observe that g is continuous at x and $\overline{D}g(x) > A$, $\underline{D}g(x) < -A$. By Lemma 4.1, we can choose a_x, b_x such that

$$x_{i-1} < a_x < x < b_x < x_i \tag{4.3}$$

and

$$\frac{g(b_x) - g(a_x)}{b_x - a_x} > A, \qquad \text{if } g(x_i) \le g(x_{i-1}),$$

$$\frac{g(b_x) - g(a_x)}{b_x - a_x} < -A, \qquad \text{if } g(x_i) \ge g(x_{i-1}).$$

It is clear that

$$G \setminus P \subseteq \bigcup_{x \in G \setminus P} (a_x, b_x).$$

By Lemma 4.4, there is a nonempty family of intervals $\{(a_x, b_x)\}_{x \in U}$, where U is a finite subset of $E \setminus P$, such that

$$\frac{1}{2} m^*(G) = \frac{1}{2} m^*(G \setminus P) \le \sum_{x \in U} (b_x - a_x).$$

By Lemma 4.5, there is a nonempty subset $V \subseteq U$ such that the intervals (a_x, b_x) with $x \in V$ are pairwise disjoint and

$$\sum_{x \in V} (b_x - a_x) \ge \frac{1}{3} \sum_{x \in U} b_x - a_x) \ge \frac{1}{6} m^*(G). \qquad (4.4)$$

Let Q be the set of endpoints of the intervals (a_x, b_x) with $x \in V$ and let $Q_i = (P \cup Q) \cap [x_{i-1}, x_i]$ for $1 \le i \le n$. Let $[x_{i-1}, x_i]$ be an interval containing at least one of the intervals (a_x, b_x) with $x \in V$ (cf. (4.3)). By applying the results of Lemma 4.2 to the function g over $[x_{i-1}, x_i]$ and the partition Q_i, we obtain

$$V_{x_{i-1}}^{x_i}(Q_i, g) > |g(x_i) - g(x_{i-1})| + A \cdot L_i,$$

where L_i is the total length of intervals $\{(a_x, b_x)\}_{x \in Q_i}$. On the other hand, for an interval $[x_{i-1}, x_i]$ that does not contain any of the intervals (a_x, b_x) with $x \in V$, we have

$$V_{x_{i-1}}^{x_i}(Q_i, g) \ge |g(x_i) - g(x_{i-1})|.$$

By adding the last two displayed inequalities for $1 \le i \le n$, we obtain

$$V_a^b(P \cup Q, g) = \sum_{i=1}^n V_{x_{i-1}}^{x_i}(Q_i, g) > V_a^b(P, g) + A \cdot \sum_{x \in V} (b_x - a_x)$$

$$> \left[V_a^b(g) - A \cdot \frac{1}{6} m^*(E) \right] + A \cdot \frac{1}{6} m^*(E) = V_a^b(g),$$

by (4.2) and (4.4). This inequality is the desired contradiction. $\qquad \square$

Theorem 4.7. *Let f be a BV-function on $[a, b]$ and let E be the set of all points in (a, b) where the upper derivative is not finite. Then E is a set of measure zero.*

Proof. We prove the claim for the case when

$$E = \{x \in (a,b) : \overline{D}f(x) = \infty\}.$$

The proof is similar for the other possible case.

Again, let E_c be the set of all points in (a,b) at which f is continuous. We need to prove that the set

$$E' = \{x \in E_c : \overline{D}f(x) - \infty\}$$

is of measure zero. The proof is by contradiction, so we suppose that $m^*(E') > 0$.

Let A be an arbitrary positive number and x be a point in the set E'. Then $\overline{D}f(x) > A$. By Lemma 4.1, we can choose a_x, b_x such that

$$a < a_x < x < b_x < b$$

and

$$\frac{g(b_x) - g(a_x)}{b_x - a_x} > A, \qquad \text{if } g(b) \leq g(a),$$

$$\frac{g(b_x) - g(a_x)}{b_x - a_x} < -A, \qquad \text{if } g(b) \geq g(a).$$

It is clear that

$$E' \subseteq \bigcup_{x \in E'} (a_x, b_x).$$

By Lemma 4.4, there is a family of intervals $\{(a_x, b_x)\}_{x \in U}$, where U is a finite subset of E', such that

$$\frac{1}{2}m^*(E') \leq \sum_{x \in U} (b_x - a_x).$$

By Lemma 4.5, there is a subset $V \subseteq U$ such that the intervals (a_x, b_x) with $x \in V$ are pairwise disjoint and

$$\sum_{x \in V} (b_x - a_x) \geq \frac{1}{3} \sum_{x \in U} (b_x - a_x) \geq \frac{1}{6}m^*(E').$$

Let P be a partition of $[a, b]$ that includes the endpoints a, b and all endpoints of the intervals (a_x, b_x) with $x \in V$. By Lemma 4.2,

$$V_a^b(P, f) > |f(b) - f(a)| + A \cdot \sum_{x \in V} (b_x - a_x) \geq A \cdot \frac{1}{6}m^*(E').$$

Because A is an arbitrary positive number and $m^*(E') > 0$, this contradicts our assumption that f is a BV-function. $\qquad\square$

By combining the results of Theorems 4.6 and 4.7 we obtain the following theorem.

Theorem 4.8. *A BV function on a closed bounded interval is differentiable almost everywhere.*

The converse is not true. For instance, the function defined by $f(x) = 1/x$ for $x \in (0, 1]$ and $f(0) = 0$ is differentiable over $(0, 1]$ but is not of the bounded variation over $[0, 1]$.

Corollary 4.3. *A monotone function on a closed interval is differentiable a.e. on the interval because it is a BV-function (cf. Example 4.7).*

Theorem 4.9. *If f is an increasing function on $[a, b]$, then its derivative f' is measurable and*

$$\int_a^b f' \leq f(b) - f(a),$$

so that f' is integrable.

Compare this result with the fundamental theorem of calculus, FTC 2 (see the preamble to this chapter).

Proof. Let us set

$$f(x) = f(b) \quad \text{for } x \in (b, b+1].$$

At every point $x \in [a, b]$ where the derivative exists, we have

$$f'(x) = \lim \frac{f(x + \frac{1}{n}) - f(x)}{\frac{1}{n}} = \lim n\Big[f\Big(x + \frac{1}{n}\Big) - f(x)\Big].$$

Recall that an increasing function on $[a, b]$ is measurable (cf. Exercise 2.46). Thus f' is the limit of a sequence of measurable functions converging a.e. By Theorem 2.33, f' is a measurable function. It is clear that it is nonnegative. Inasmuch as f is an increasing function, the integral

$$\int_a^b f\Big(x + \frac{1}{n}\Big)\, dx$$

can be taken in the Riemann sense (cf. Exercise 3.6). Therefore the change of variable can be used to obtain

$$\int_a^b f\Big(x + \frac{1}{n}\Big)\, dx = \int_{a+1/n}^{b+1/n} f(x)\, dx.$$

By Fatou's Lemma (Theorem 3.16),

$$\int_a^b f'(x)\, dx \leq \liminf \int_a^b n\Big[f\Big(x + \frac{1}{n}\Big) - f(x)\Big]\, dx$$

$$= \liminf \Big[n \int_a^b f\Big(x + \frac{1}{n}\Big)\, dx - n \int_a^b f(x)\, dx\Big]$$

$$\liminf \Big[n \int_{a+1/n}^{b+1/n} f - n \int_a^b f\Big]$$

$$= \liminf \Big[n \int_b^{b+1/n} f - n \int_a^{a+1/n} f\Big].$$

We have $f(x) = f(b)$ over the intervals $[b, b + 1/n]$ and $f(x) \geq f(a)$ over the intervals $[a, 1/n]$. Therefore,

$$n \int_b^{b+1/n} f = f(b) \quad \text{and} \quad n \int_a^{a+1/n} f \geq f(a),$$

so $\int_a^b f'(x)\, dx \leq f(b) - f(a)$, which is the desired result. □

Example 4.9. Let $c(x)$ be the Cantor function on $[0, 1]$. Then $c'(x) = 0$ a.e. on $[0, 1]$ (cf. Example 2.3), so we have the strict inequality

$$0 = \int_0^1 c' < c(1) - c(0) = 1$$

in Theorem 4.9.

4.4 Differentiation of an Indefinite Integral

Let f be an integrable function on the closed interval $[a, b]$. The *indefinite integral* of f is the function F defined on $[a, b]$ by

$$F(x) = \int_a^x f + C,$$

for every choice of the constant C. It is clear that $F(a) = C$, so we can write

$$F(x) = \int_a^x f + F(a)$$

for any particular indefinite integral F of f.

In this section we prove the following theorem which is a version of the fundamental theorem of calculus, FTC 1 (cf. the preamble to this chapter):

Theorem 4.10. *Let f be an integrable function on $[a, b]$ and us suppose that*

$$F(x) = \int_a^x f + F(a).$$

Then $F'(x) = f(x)$ for almost all x in $[a, b]$.

Before establishing the claim of this theorem, we prove several lemmas.

Lemma 4.7. *The indefinite integral F of an integrable function f on the interval $[a, b]$ is a continuous function on this interval.*

Proof. By Theorem 3.19, given $\varepsilon > 0$, there is $\delta_1 > 0$ such that $\int_A |f| < \varepsilon$ for any subset A of $[a, b]$ with $m(A) < \delta_1$. (Recall (cf. Theorem 3.20) that $|f|$ is integrable if and only if f is integrable.) For a given $x \in [a, b]$, we choose

$$A = (x - \delta_1/3, x + \delta_1/3) \cap [a, b].$$

Clearly, $m(A) < \delta_1$. Therefore,

$$|F(y) - F(x)| = \left| \int_x^y f \right| \leq \left| \int_x^y |f| \right| \leq \int_A |f| < \varepsilon,$$

for any $y \in [a, b]$ such that $|y - x| < \delta = \delta_1/3$. Thus, F is continuous at x. $\square$

Lemma 4.8. *The indefinite integral F of an integrable function f on the interval $[a, b]$ is a BV-function on this interval.*

Proof. For a partition

$$a = x_0 < x_1 \cdots < x_n = b,$$

we have

$$\sum_{i=1}^n |F(x_i) - F(x_{i-1})| = \sum_{i=1}^n \left| \int_{x_{i-1}}^{x_i} f \right| \leq \sum_{i=1}^n \int_{x_{i-1}}^{x_i} |f| = \int_a^b |f|.$$

Therefore, $V_a^b(F) \leq \int_a^b |f| < \infty$. $\square$

Lemma 4.9. *Let f be a positive measurable function on a bounded set E of positive measure. Then*

$$\int_E f > 0.$$

Proof. Let $E_n = \{x \in E : f(x) > 1/n\}$. Then

$$E_1 \subseteq E_2 \subseteq \cdots \subseteq E_n \cdots \subseteq E \quad \text{and} \quad E = \bigcup_{i=1}^\infty E_i.$$

By Theorem 2.24,

$$\lim m(E_n) = m(E) > 0.$$

Hence, there is n such that $m(E_n) > 0$. Then, by Theorem 3.2, we have

$$\int_E f = \int_{E_n} f + \int_{E \backslash E_n} f \geq \int_{E_n} f \geq \frac{1}{n} m(E_n) > 0,$$

which is the desired result. $\square$

Lemma 4.10. (Cf. Exercise 3.9.) *Let f be an integrable function on $[a, b]$. If*

$$\int_a^x f = 0 \qquad \text{for all } x \in [a, b],$$

then $f(x) = 0$ a.e. on $[a, b]$.

Proof. The proof is by contradiction. Let us represent the set

$$\{x \in [a, b] : f(x) \neq 0\}$$

as the union

$$\{x \in [a, b] : f(x) > 0\} \cup \{x \in [a, b] : f(x) < 0\}$$

and assume that $m(E) > 0$, where

$$E = \{x \in [a, b] : f(x) > 0\}$$

(otherwise, consider the set where f is negative). Then $m_*(E) > 0$ and, by the definition of inner measure, there is a closed set $F \subseteq E$ with $m(F) > 0$. By Lemma 4.9,

$$\int_F f > 0.$$

The set $G = (a, b) \setminus F$ is open and therefore is the union of an at most countable family of pairwise disjoint open intervals,

$$G = \bigcup_{i \in J} (a_i, b_i).$$

We have

$$\int_G f = \int_a^b f - \int_F f = - \int_F f \neq 0.$$

Inasmuch as

$$\int_G f = \sum_{i \in J} \int_{a_i}^{b_i} f,$$

there is $i \in J$ such that $\int_{a_i}^{b_i} f \neq 0$. On the other hand,

$$\int_{a_i}^{b_i} f = \int_a^{b_i} f - \int_a^{a_i} f = 0.$$

This contradiction completes the proof. $\qquad\qquad\qquad\qquad\qquad\square$

Lemma 4.11. *Let f be a bounded measurable function on $[a, b]$, and let*

$$F(x) = \int_a^x f + F(a).$$

Then $F'(x) = f(x)$ for almost all $x \in [a, b]$.

Proof. By Lemma 4.8 and Theorem 4.8, the derivative $F'(x)$ exists a.e. on $[a, b]$. Let us set

$$f(x) = f(b) \quad \text{for } x \in (b, b+1]$$

and define (cf. proof of Theorem 4.9)

$$f_n(x) = \frac{F(x + \frac{1}{n}) - F(x)}{\frac{1}{n}} = n\left[\int_a^{x+1/n} f - \int_a^x f\right] = n\int_x^{x+1/n} f.$$

Because f is bounded, there is $M > 0$ such that $|f| \leq M$ on $[a, b]$. It follows that

$$|f_n(x)| = n\left|\int_x^{x+1/n} f\right| \leq n\int_x^{x+1/n} |f| \leq M, \qquad \text{for all } x \in [a, b].$$

Because $f_n(x)$ converges to $F'(x)$ pointwise a.e. on $[a, b]$, the bounded convergence theorem (Theorem 3.10) implies

$$\int_a^c F' = \lim \int_a^c f_n = \lim n \int_a^c \left[F\left(x + \frac{1}{n}\right) - F(x)\right] dx$$

$$= \lim n\left[\int_a^c F\left(x + \frac{1}{n}\right) dx - \int_a^c F(x)\, dx\right]$$

$$= \lim n\left[\int_{a+1/n}^{c+1/n} F - \int_a^c F\right]$$

$$= \lim \left[n\int_c^{c+1/n} F - n\int_a^{a+1/n} F\right] = F(c) - F(a) = \int_a^c f,$$

for every $c \in [a, b]$. The middle equality in the last line holds by continuity of F (cf. Lemma 4.7). Therefore,

$$\int_a^c (F' - f) = 0, \qquad \text{for all } c \in [a, b].$$

By Lemma 4.10, $F' = f$ a.e. on $[a, b]$.   □

Lemma 4.12. *Let f be a nonnegative integrable function on $[a, b]$, and let*

$$F(x) = \int_a^x f.$$

Then $F'(x) = f(x)$ for almost all $x \in [a, b]$.

Proof. Let n be a fixed natural number. We define

$$f_n(x) = \min\{f(x), n\}, \qquad \text{for } x \in [a, b].$$

Because $f(x) - f_n(x) \geq 0$ on $[a, b]$, the function

$$G_n(x) = \int_a^x (f - f_n)$$

is an increasing function of x. By Theorem 4.8, G_n' exists a.e. on $[a, b]$. It is clear that $G_n'(x) \geq 0$ for all points x where it is defined, because G_n is an increasing function. By Lemma 4.11,

$$\frac{d}{dx} \int_a^x f_n = f_n(x) \quad \text{a.e. on } [a, b].$$

Therefore,

$$F'(x) = \frac{d}{dx} \int_a^x f = \frac{d}{dx} \left[G_n(x) + \int_a^x f_n \right] \geq f_n(x) \quad \text{a.e. on } [a, b].$$

Because n is an arbitrary natural number, $F'(x) \geq f(x)$ a.e. on $[a, b]$. Consequently,

$$\int_a^b F' \geq \int_a^b f = F(b) - F(a).$$

Observe that F is an increasing function, for $f(x) \geq 0$ on $[a, b]$. By Theorem 4.9,

$$\int_a^b F' \leq F(b) - F(a).$$

Hence,

$$\int_a^b F' = F(b) - F(a) = \int_a^b f,$$

so $\int_a^b (F' - f) = 0$. Inasmuch as $F'(x) \geq f(x)$, by Lemma 4.11, we have $F'(x) = f(x)$ a.e. on $[a, b]$. □

Let f be an integrable function on $[a, b]$. By Lemma 4.12, the indefinite integrals of the functions f^+ and f^- are differentiable a.e. and their derivatives equal to the respective functions f^+ and f^-. This proves the claim of Theorem 4.10.

4.5 Absolutely Continuous Functions

Definition 4.2. *A real-valued function f defined on $[a, b]$ is said to be absolutely continuous on $[a, b]$ if, given $\varepsilon > 0$, there is $\delta > 0$ such that*

$$\sum_{i=1}^n |f(b_i) - f(a_i)| < \varepsilon$$

for every finite set $\{(a_i, b_i) : 1 \le i \le n\}$ *of pairwise disjoint intervals with*

$$\sum_{i=1}^{n} (b_i - a_i) < \delta.$$

An absolutely continuous function is also continuous; just take $n = 1$ in the above definition. However, the converse does not hold. For instance, the Cantor function c (cf. Example 2.3) is continuous but not absolutely continuous. Indeed, in the kth step of the construction of the Cantor set, we obtained the set C_k which is the union of 2^k closed intervals (a_i, b_i), $1 \le i \le 2^k$, each of which has length $(1/3)^k$ (cf. Example 1.2). We have

$$\sum_{i=1}^{2^k} (b_i - a_i) = (2/3)^k, \quad \text{while} \quad \sum_{i=1}^{2^k} [c(b_i) - c(a_i)] = 1,$$

because the Cantor function is constant on each of the intervals that comprise the relative complement $[0, 1] \setminus C_k$. Therefore, for $\varepsilon = 1$, we must have $\delta < (2/3)^k$ for all $k \in \mathbb{N}$ which is impossible.

It is not difficult to show that linear combinations of absolutely continuous functions are absolutely continuous (cf. Exercise 4.22).

Let f be a BV-function on a closed interval $[\alpha, \beta]$, and let P be a partition of $[\alpha, \beta]$ defined by the sequence of points

$$\gamma_0 = \alpha < \gamma_1 < \cdots < \gamma_n = \beta.$$

Consider the family of open intervals $\{(\gamma_i, \gamma_{i-1})\}_{1 \le i \le n}$. Then

$$\sum_{i=1}^{n} |f(\gamma_i) - f(\gamma_{i-1})| = V_\alpha^\beta(P, f)$$

and

$$\sum_{i=1}^{n} (\gamma_i - \gamma_{i-1}) = \beta - \alpha.$$

We will use this observation implicitly in the rest of this section.

Theorem 4.11. *An absolutely continuous function f on $[a, b]$ is of bounded variation.*

Proof. Let $\delta > 0$ be the response to the $\varepsilon = 1$ challenge in Definition 4.2. Let us partition the interval $[a, b]$ into N intervals by points

$$a = x_0 < x_1 < \cdots < x_N = b$$

in such a way that the length of each interval is less than δ. By the definition of absolute continuity, $V_{x_{i-1}}^{x_i}(f) \le 1$, for $1 \le i \le N$. The additivity property of the total variation (cf. Theorem 4.3) implies

$$V_a^b(f) = \sum_{i=1}^{N} V_{x_{i-1}}^{x_i}(f) \le N.$$

Therefore f is a BV-function.   □

Theorem 4.12. *A function f is absolutely continuous if and only if it is equal to the difference of two increasing absolutely continuous functions.*

Proof. (Necessity.) Suppose that f is an absolutely continuous function on the interval $[a, b]$. By Theorem 4.11, f is a BV-function. By Corollary 4.2,

$$f = V_a^x(f) - (V_a^x(f) - f),$$

where functions $V_a^x(f)$ and $f - V_a^x(f)$ are both increasing functions on $[a, b]$. Thus, it suffices to show that the function $V_a^x(f)$ is absolutely continuous.

Given $\varepsilon > 0$, let $\delta > 0$ be the response to $\varepsilon/2$ according to Definition 4.2 for the function f, and let $\{(a_i, b_i) : 1 \le i \le n\}$ be a family of pairwise disjoint subintervals of $[a, b]$ with $\sum_{i=1}^{n}(b_i - a_i) < \delta$. We have

$$\sum_{i=1}^{n} |V_a^{b_i}(f) - V_a^{a_i}(f)| = \sum_{i=1}^{n} V_{a_i}^{b_i}(f) = \sup \sum_{i=1}^{n} V_{a_i}^{b_i}(P_i, f),$$

where the supremum is taken over partitions P_i of the intervals (a_i, b_i) for all $1 \le i \le n$. The set $\bigcup_{i=1}^{n} P_i$ defines a partition of the union $\bigcup_{i=1}^{n}(a_i, b_i)$ into subintervals of total length less than δ. Therefore,

$$\sum_{i=1}^{n} V_{a_i}^{b_i}(P_i, f) < \frac{\varepsilon}{2},$$

by our choice of δ. By taking the supremum on the left side, we obtain

$$\sum_{i=1}^{n} |V_a^{b_i}(f) - V_a^{a_i}(f)| = \sup \sum_{i=1}^{n} V_{a_i}^{b_i}(P_i, f) \le \frac{\varepsilon}{2} < \varepsilon.$$

It follows that $V_a^x(f)$ is an absolutely continuous function.

(Sufficiency.) The proof is left as Exercise 4.22.   □

Theorem 4.13. *The indefinite integral F of an integrable function f on $[a, b]$,*

$$F(x) = \int_a^x f + F(a),$$

is an absolutely continuous function on $[a, b]$.

Proof. Let $\{(a_i, b_i) : 1 \leq i \leq n\}$ be a set of pairwise disjoint subintervals of $[a, b]$, and let $G = \bigcup_{i=1}^{n}(a_i, b_i)$. Then

$$\sum_{i=1}^{n} |F(b_i) - F(a_i)| = \sum_{i=1}^{n}\left|\int_{a_i}^{b_i} f\right| \leq \sum_{i=1}^{n}\int_{a_i}^{b_i}|f| = \int_{G}|f|.$$

By Theorem 3.19, for a given $\varepsilon > 0$ there is $\delta > 0$ such that $\int_{G}|f| < \varepsilon$ provided that $m(G) = \sum_{i=1}^{n}(b_i - a_i) < \delta$. It follows that F is an absolutely continuous function. $\square$

In the rest of this section we prove Lebesgue's version of the fundamental theorem of calculus, FTC 2 (Theorem 4.14). First, we prove four lemmas.

Let g be a continuous function on $[a, b]$. A point $x \in [a, b]$ is called a *shadow point* of g if there is a point $z > x$ in $[a, b]$ such that $g(z) > g(x)$. This concept is illustrated by the drawing in Fig. 4.1. The shadow points correspond to points in the "valleys" of the graph. Arrows are rays of light coming from the "sun rising in the east."

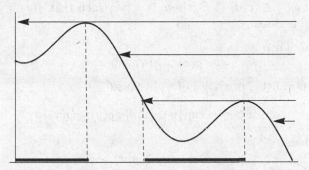

Figure 4.1. Shadow points

Lemma 4.13. *The set E of shadow points of g in (a, b) is open.*

Proof. We may assume that $E \neq \varnothing$. Let x be a shadow point in (a, b) so $g(z) > g(x)$ for some $z \in (x, b]$. Inasmuch as g is a continuous function, for $\varepsilon = g(z) - g(x)$, there is $\delta > 0$ such that

$$|g(y) - g(x)| < \varepsilon, \qquad \text{for } y \in [a, b] \text{ such that } |y - x| < \delta.$$

Therefore,

$$g(y) < g(x) + \varepsilon = g(z), \qquad \text{for } y \in (x - \delta, x + \delta) \cap (a, b),$$

that is, all points in the interval $(x - \delta, x + \delta) \cap (a, b)$ are shadow points. It follows that E is an open set. $\square$

Lemma 4.14. *Suppose that the set E of shadow points of g in (a,b) is not empty and let (c,d) be a component interval of E. Then*

$$g(c) \leq g(d).$$

Proof. We will show that $g(x) \leq g(d)$ for all $x \in (c,d)$. Then the desired result follows from the continuity of g.

For a given $x \in (c,d)$, we define

$$\gamma = \sup\{y \in [x,d] : g(y) \geq g(x)\}.$$

Note that γ is well defined because $x \in \{y \in [x,d] : g(y) \geq g(x)\}$. Because g is continuous, $g(\gamma) \geq g(x)$. Hence, if $\gamma = d$, we are done.

Suppose that $\gamma < d$. Then

$$g(d) < g(x),$$

because otherwise, $d \in \{y \in [x,d] : g(y) \geq g(x)\}$, which is false by the definition of γ.

Inasmuch as $\gamma \in (c,d) \subseteq E$, there is $z > \gamma$ such that $g(z) > g(\gamma)$. There are two possible cases:

1. $\gamma < z \leq d$. Then

$$g(z) > g(\gamma) \geq g(x),$$

which contradicts our choice of γ, because

$$z > \gamma = \sup\{y \in [x,d] : g(y) \geq g(x)\}.$$

2. $z > d$. Then

$$g(z) > g(\gamma) \geq g(x) > g(d).$$

It follows that d is a shadow point which is false because $d \notin E$.

These contradictions show that $\gamma = d$ and hence $g(x) \leq g(d)$ for all $x \in (c,d)$. $\square$

Lemma 4.15. *Let f be an increasing absolutely continuous function on $[a,b]$. If Z is a subset of $[a,b]$ of measure zero, then $f(Z)$ is also a set of measure zero.*

Proof. It is not difficult to see that we may assume that Z is a subset of the open interval (a,b).

Given $\varepsilon > 0$, let $\delta > 0$ be as in Definition 4.2. Because $m(Z) = 0$, there is an open set $G' \supseteq Z$ with $m(G') < \delta$. Consider the open set $G = G' \cap (a,b)$. The set G is the union of its component intervals,

$$G = \bigcup_{i \in J} (a_i, b_i),$$

so $\sum_{i \in J}(b_i - a_i) < \delta$, because $G \subseteq G'$. Inasmuch as f is an increasing function,

$$f(G) = \bigcup_{i \in J}(f(a_i), f(b_i)) \subseteq (f(a), f(b)),$$

where $\{(f(a_i), f(b_i))\}_{i \in J}$ is a family of pairwise disjoint intervals.

We have

$$m(f(G)) = \sum_{i \in J}(f(b_i) - f(a_i)) < \varepsilon$$

(cf. Exercise 4.26). Therefore, $m(f(G)) = 0$, because ε is an arbitrary positive number. Since $f(Z) \subseteq f(G)$, we have $m(f(Z)) = 0$. $\square$

Lemma 4.16. *If f is an increasing absolutely continuous function on $[a, b]$ with $f'(x) = 0$ a.e. on (a, b), then f is a constant function.*

Proof. Let $E = \{x \in (a, b) : f'(x) = 0\}$. Because $f'(x) = 0$ a.e. on (a, b), the set $Z = [a, b] \setminus E$ is of measure zero. We have

$$f([a, b]) = [f(a), f(b)],$$

because f is an increasing function. Therefore,

$$f(b) - f(a) = m(f(Z)) + m(f(E)) = m(f(E)),$$

by Lemma 4.15. To prove the claim of the lemma, it suffices to show that $m(f(E)) = 0$.

Let ε be a positive number. For each $x \in E$, select $z > x$ in E such that

$$\frac{f(z) - f(x)}{z - x} < \varepsilon,$$

or, equivalently,

$$\varepsilon x - f(x) < \varepsilon z - f(z).$$

Thus the set E is a subset of the set F of shadow points of the function $g(x) = \varepsilon x - f(x)$. We may assume that $E \neq \varnothing$. (Otherwise, $m(f(E)) = 0$.) Then, by Lemma 4.13, the set F is the union of open intervals

$$F = \bigcup_{i \in J}(c_i, d_i).$$

By Lemma 4.14, $g(c_i) \leq g(d_i)$ for $i \in J$. Therefore,

$$f(d_i) - f(c_i) \leq \varepsilon(d_i - c_i), \qquad \text{for } i \in J.$$

Because f is an increasing function, $f(F)$ is the union of pairwise disjoint intervals $(f(d_i), f(c_i))$ (here, we assume that $(\alpha, \alpha) = \varnothing$) and hence

$$m(f(F)) = \sum_{i \in J}(f(d_i) - f(c_i)) \leq \varepsilon \sum_{i \in J}(d_i - c_i) \leq \varepsilon(b - a).$$

Since ε is an arbitrary positive number, $f(F)$ is a set of measure zero. Because $E \subseteq F$, the set $f(E)$ is also of measure zero. $\qquad \square$

Theorem 4.14. *The derivative $f = F'$ of an absolutely continuous function F on $[a, b]$ is integrable on $[a, b]$ and*

$$\int_a^x f = F(x) - F(a).$$

Proof. By Theorem 4.12, we may assume that F is an increasing function. By Theorem 4.9, f is integrable and

$$\int_u^v f \leq F(u) - F(v) \qquad \text{for all } a \leq u < v \leq b.$$

The function $G(x) = F(x) - \int_a^x f$ is also increasing because

$$G(v) - G(u) = F(v) - F(v) - \int_u^v f \geq 0.$$

for $a \leq u < v \leq b$. Moreover, the function G is absolutely continuous as the difference of two absolutely continuous functions (cf. Theorem 4.13). By Theorem 4.10, we have

$$G'(x) = F'(x) - f(x) = 0 \quad \text{a.e. on } [a, b].$$

Hence, by Lemma 4.16, G is a constant function, $G(x) = C$ on $[a, b]$. Thus,

$$C = F(x) - \int_a^x f,$$

and for $x = a$, we obtain the desired result. $\qquad \square$

Notes

In 1872, Karl Weierstrass presented to the Berlin Academy an example of continuous but nowhere differentiable function:

$$\sum_{k=0}^{\infty} b^k \cos(a^k \pi x),$$

where $0 < b < 1$ and a is an odd integer for which $ab > 1 + 3\pi/2$. He believed that it was a matter of time before someone found an example of a monotonic function with these properties. However, in 1904 Henri Lobesgue proved that any continuous monotonic function is differentiable almost everywhere. In 1910, George Faber demonstrated that continuity is not necessary, thus extending Lebesgue's result to the functions of bounded variation (cf. Theorem 4.8). It is a common belief that Lebesgue's theorem is "one of the most striking and most important in real variable theory," as Frigyes Riesz and Béla Sz.-Nagy remark in their book [RSN90]. In our exposition of this topic we follow an approach presented in [Aus65] and [Bot03].

Lemma 4.13 is known as Riesz's Rising Sun Lemma in the pertinent literature. It also has applications in functional and harmonic analyses.

Exercises

4.1. Describe the punctured balls $B'_{0.1}$, $B'_{1.0}$, $B'_{1.1}$, $B'_{2.0}$, and B'_{137} centered at $x = 1$ in the interval $[0, 3]$.

4.2. Prove that $\underline{D}f(x) = -\overline{D}[-f(x)]$.

4.3. Show that the function

$$f(x) = \begin{cases} x^2 \sin \frac{1}{x^2}, & \text{if } x \neq 0, \\ 0, & \text{if } x = 0 \end{cases}$$

is not of a bounded variation over the interval $[-1, 1]$.

4.4. Show that the function

$$f(x) = \begin{cases} x^2 \cos \frac{1}{x}, & \text{if } x \neq 0, \\ 0, & \text{if } x = 0 \end{cases}$$

is a BV-function over the interval $[0, 1]$.

4.5. Let f be a BV-function on $[a, b]$ and $[c, d]$ be a nontrivial subinterval of $[a, b]$. Show that the restriction of f to $[c, d]$ is a BV-function over $[c, d]$.

4.6. Show that the sum, difference, and product of two BV-functions is a BV-function.

4.7. Let f and g be two BV-functions on $[a, b]$. Show that f/g is a BV-function provided that $g(x) \geq \sigma > 0$ on $[a, b]$.

4.8. Show that Dirichlet's function

$$f(x) = \begin{cases} 1, & \text{if } x \text{ is a rational number,} \\ 0, & \text{otherwise} \end{cases} \quad \text{on } [0, 1]$$

is not of a bounded variation.

4.9. Show that if f has a bounded derivative on $[a, b]$, then f is of bounded variation.

4.10. Show that $V_a^b(f) = f(b) - f(a)$ if and only if f is an increasing function on $[a, b]$.

4.11. Show that a BV-function is bounded.

4.12. Show that if f is a BV-function on $[a, b]$, then so is $|f|^p$, $p \geq 1$.

4.13. Let F be a closed subset of $[a, b]$. Show that there is a continuous function f on $[a, b]$ such that $\{x \in [a, b] : f(x) = 0\} = F$.

4.14. Prove that if f is a continuous function on $[a, b]$ and $|f|$ is of bounded variation on $[a, b]$, then so is f. Show that continuity is an essential assumption.

4.15. If f and g are BV-functions on $[a, b]$, then so is

$$h(x) = \max\{f(x), g(x)\}.$$

4.16. Let f and g be BV-functions and c be an arbitrary constant. Show that

$$V_a^b(f + g) \leq V_a^b(f) + V_a^b(g) \quad \text{and} \quad V_a^b(cf) = |c| V_a^b(f).$$

Find f and g for which the first inequality is strict.

4.17. Let f be a BV-function. Show that if $G(x) = V_a^x(f)$ is continuous, then so is f.

4.18. Let x be a point in the open interval I. Show that there is an open interval $J \subseteq I$ with rational endpoints that contains x.

4.19. Let $\{I_1, \ldots, I_n\}$ be a set of bounded intervals such that $m(I_1) \geq m(I_i)$ and $I_i \cap I_1 \neq \varnothing$ for all $1 \leq i \leq n$. Show that

(a) $\bigcup_{i=1}^n I_i$ is a bounded interval.
(b) $m(I_1) \geq \frac{1}{3} m\left(\bigcup_{i=1}^n I_i\right)$.

4.20. Show that a bounded set E that is not of measure zero has a positive outer measure, $m^*(E) > 0$.

4.21. Let A be a finite or countable subset of a bounded set E. Show that E is a set of measure zero if and only if $E \setminus A$ is a set of measure zero.

4.22. Let f and g be absolutely continuous functions on (a, b) and k be an arbitrary constant. Prove that the functions $f + g$, kf, and fg are absolutely continuous.

4.23. Give an example of a continuous function g on $[a, b]$ for which we have a strict inequality for a component interval in Lemma 4.14.

4.24. Give an example of a continuous increasing function f on $[0, 1]$ such that $m(f(Z)) \neq 0$ for some set $Z \subseteq [0, 1]$ of measure zero.

4.25. Prove that the function f defined by

$$f(x) = \begin{cases} x \cos \frac{\pi}{x}, & \text{if } x \in (0, 1], \\ 0, & \text{if } x = 0, \end{cases}$$

is continuous but not absolutely continuous.

4.26. Let f be an absolutely continuous function on $[a, b]$. Prove that for every $\varepsilon > 0$ there is $\delta > 0$ such that

$$\sum_{i=1}^{\infty} |f(b_i) - f(a_i)| < \varepsilon$$

for any countable set of pairwise disjoint intervals $\{(a_i, b_i) : i \in \mathbb{N}\}$ with the total length less that δ.

4.27. Let f and g be integrable functions on $[a, b]$, and let

$$F(x) = F(a) + \int_a^x f, \qquad G(x) = G(a) + \int_a^x g.$$

Prove that

$$\int_a^b (G \cdot f) + \int_a^b (g \cdot F) = F(b)G(b) - F(a)G(a).$$

4.28. (Integration by parts) If f and g are absolutely continuous functions on $[a, b]$, then

$$\int_a^b (f \cdot g') + \int_a^b (f' \cdot g) = f(b)g(b) - f(a)g(a).$$

A

Measure and Integral over Unbounded Sets

As presented in Chaps. 2 and 3, Lebesgue's theory of measure and integral is limited to functions defined over bounded sets. There are several ways of introducing the theory for arbitrary domains (and even for functions with values in the set of extended reals). However, instead of developing a general theory from scratch, in this Appendix we lay out an approach utilizing properties of measure and integral that were established in the main text for functions over bounded domains.

A.1 The Measure of an Arbitrary Set

In this section we extend the concept of measurability from bounded to arbitrary sets of real numbers.

Definition A.1. *A set E of real numbers is said to be* measurable *if all the sets*
$$E \cap [-n, n], \quad n \in \mathbb{N}$$
are measurable in the sense of Definition 2.5.

It is clear that a bounded set is measurable in the sense of Definition 2.5 if and only if it is measurable in the sense of Definition A.1.

Another possible way of introducing the concept of measurability is to define a set E to be measurable if the intersection $E \cap A$ of this set with an arbitrary bounded measurable set A is measurable. However, this approach produces the same class of measurable sets as defined above.

Theorem A.1. *A set E is measurable in the sense of Definition A.1 if and only if, for any bounded measurable set A, the set $E \cap A$ is measurable in the sense of Definition 2.5.*

S. Ovchinnikov, *Measure, Integral, Derivative: A Course on Lebesgue's Theory,*   129
Universitext, DOI 10.1007/978-1-4614-7196-7,
© Springer Science+Business Media New York 2013

Proof. Because the sufficiency part is trivial, we proceed with the necessity part. Let us assume that the set E is measurable in the sense of Definition A.1 and let A be a bounded measurable set of real numbers. Inasmuch as A is bounded, there is $n \in \mathbb{N}$ such that $A \subseteq [-n, n]$. Then

$$E \cap A = (E \cap [-n, n]) \setminus ([-n, n] \setminus A).$$

The sets $E \cap [-n, n]$, $[-n, n]$, and A are bounded and measurable. It follows that the set $E \cap A$ is bounded and measurable. $\square$

Definition A.2. *Let E be a measurable set of real numbers. The measure $m(E)$ is defined by*

$$m(E) = \sup\{m(E \cap [-n, n]) : n \in \mathbb{N}\}.$$

Because

$$E \cap [-1, 1] \subseteq E \cap [-2, 2] \subseteq \cdots \subseteq E \cap [-n, n] \subseteq \cdots,$$

the sequence of numbers $(m(E \cap [-n, n]))$ is increasing. If it is bounded, then it is convergent (cf. Theorem 1.3) and

$$m(E) = \lim m(E \cap [-n, n]).$$

Otherwise, we set $m(E) = \infty$ (see conventions in Sect. 1.2). Thus $m(E)$ is defined for every measurable set and assumes its values in the set of extended real numbers (recall that ∞ stands for $+\infty$). By convention, $a + \infty = \infty$ and $a \leq \infty$ for all extended real numbers a. We also make the convention that $\lim a_n = \infty$ if (a_n) is an increasing sequence of extended real numbers with $a_k = \infty$ for some $k \in \mathbb{N}$.

Example A.1. 1. Let I be an unbounded interval. By choosing a sufficiently large n, we can make the length of the interval $I \cap [-n, n]$ larger than any given number. Therefore, $m(I) = \infty$. In particular, $m(\mathbb{R}) = \infty$.
  2. Let E be a finite or countable set of points. Then

$$m(E \cap [-n, n]) = 0, \quad \text{for every } n \in \mathbb{N}.$$

It follows that $m(E) = 0$.
  3. Let

$$E = \bigcup_{k=2}^{\infty} \left[k - \frac{1}{k}, k + \frac{1}{k} \right].$$

Then

$$m(E) = \lim m(E \cap [-n, n]) = \lim \left(\sum_{k=2}^{n-1} \frac{2}{k} + \frac{1}{n} \right) = \infty.$$

4. Let $E = \bigcup_{k=1}^{\infty}[k - 2^{-k}, k + 2^{-k}]$. It is easy to verify that $m(E) - \sum_{k=1}^{\infty} 2^{-k+1} = 2$.

In what follows we present generalizations of the following properties of bounded measurable sets:

1. Open and closed sets are measurable (Theorem 2.16).
2. Countable additivity (Theorem 2.18).
3. The union and intersection of a countable family of measurable sets is measurable (Theorems 2.22 and 2.23).
4. The "continuity" properties of Lebesgue's measure (Theorems 2.24 and 2.25).

Some other properties are found in Exercises A.2–A.5.

Theorem A.2. *Any open or closed set of real numbers is measurable.*

Proof. Let E be an open subset of $\mathbb{R}$. Because E is a union of an at most countable family of pairwise disjoint open intervals (cf. Theorem 1.7), its intersection with any interval $[-n, n]$, $n \in \mathbb{N}$, is the union of an at most countable family of pairwise disjoint intervals. The latter set is measurable by Theorem 2.18.

Now let E be a closed subset of $\mathbb{R}$. Then, for every $n \in \mathbb{N}$, the intersection $E \cap [-n, n]$ is a bounded closed set and therefore is measurable. $\qquad\square$

We establish the additivity property of measure separately for finite and countable families of sets.

Theorem A.3. *Let $\{E_i\}_{i \in J}$ be a finite family of pairwise disjoint measurable sets. Then*

$$m\Big(\bigcup_{i \in J} E_i\Big) = \sum_{i \in J} m(E_i).$$

Proof. By Definition A.2 and Theorem 2.18, we have

$$m\Big(\bigcup_{i \in J} E_i\Big) = \lim m\Big(\Big(\bigcup_{i \in J} E_i\Big) \cap [-n, n]\Big)$$

$$= \lim m\Big(\bigcup_{i \in J} \Big(E_i \cap [-n, n]\Big)\Big)$$

$$= \lim \sum_{i \in J} m(E_i \cap [-n, n])$$

$$= \sum_{i \in J} \lim m(E_i \cap [-n, n])$$

$$= \sum_{i \in J} m(E_i),$$

because J is a finite set (cf. Exercise A.1). $\qquad\square$

Theorem A.4. *Let* $\{E_i\}_{i\in J}$ *be a countable family of pairwise disjoint measurable sets. Then*

$$m\Big(\bigcup_{i\in J} E_i\Big) = \sum_{i\in J} m(E_i).$$

Proof. First, we assume that $\{m(E_i)\}_{i\in J}$ is a summable family. Then, given $\varepsilon > 0$, there is a finite set $J_0 \subseteq J$ such that

$$\sum_{i\in J_0} m(E_i) > \sum_{i\in J} m(E_i) - \varepsilon.$$

Inasmuch as

$$\bigcup_{i\in J} E_i = \Big(\bigcup_{i\in J_0} E_i\Big) \cup \Big(\bigcup_{i\in J\setminus J_0} E_i\Big)$$

we have, by Theorem A.3,

$$m\Big(\bigcup_{i\in J} E_i\Big) = \sum_{i\in J_0} m(E_i) + m\Big(\bigcup_{i\in J\setminus J_0} E_i\Big).$$

Hence,

$$m\Big(\bigcup_{i\in J} E_i\Big) > \sum_{i\in J} m(E_i) - \varepsilon + m\Big(\bigcup_{i\in J\setminus J_0} E_i\Big).$$

Because ε is an arbitrary positive number, we conclude that

$$m\Big(\bigcup_{i\in J} E_i\Big) \geq \sum_{i\in J} m(E_i).$$

To prove the reverse inequality, we observe that, by the definition of m,

$$m(E_i \cap [-n, n]) \leq m(E_i),$$

for every $i, n \in \mathbb{N}$. Because $\{m(E_i)\}_{i\in J}$ is a summable family, we have, by Theorem 1.15,

$$\sum_{i\in J} m(E_i \cap [-n, n]) \leq \sum_{i\in J} m(E_i).$$

By Theorem 2.18,

$$m\Big(\Big(\bigcup_{i\in J} E_i\Big) \cap [-n, n]\Big) = \sum_{i\in J} m(E_i \cap [-n, n]).$$

It follows that

$$m\Big(\Big(\bigcup_{i\in J} E_i\Big) \cap [-n, n]\Big) \leq \sum_{i\in J} m(E_i).$$

By taking the limit as $n \to \infty$, we obtain the desired inequality:

$$m\Big(\bigcup_{i\in J} E_i\Big) \leq \sum_{i\in J} m(E_i).$$

It remains to consider the case when $\sum_{i\in J} m(E_i) = \infty$. Then, for any given real number M, there is a finite subset $J_0 \subseteq J$ such that

$$\sum_{i\in J_0} m(E_i) \geq M.$$

By Theorem A.3,

$$m\Big(\bigcup_{i\in J} E_i\Big) = \sum_{i\in J_0} m(E_i) + m\Big(\bigcup_{i\in J\setminus J_0} E_i\Big) \geq M.$$

It follows that $m\Big(\bigcup_{i\in J} E_i\Big) = \infty = \sum_{i\in J} m(E_i)$. $\square$

The proof of the following theorem is straightforward and left as an exercise (cf. Exercise A.6).

Theorem A.5. *Let* $\{E_i\}_{i\in J}$ *be a finite or countable family of measurable sets. Then*

(i) *The union* $\bigcup_{i\in J} E_i$ *is measurable.*
(ii) *The intersection* $\bigcap_{i\in J} E_i$ *is measurable.*

Finally, we establish two "continuity" properties of the extended measure.

Theorem A.6. *Let* (E_n) *be a sequence of measurable sets such that*

$$E_1 \subseteq E_2 \subseteq \cdots \subseteq E_n \subseteq \cdots$$

and let $E = \bigcup_{i=1}^{\infty} E_i$. *Then*

$$m(E) = \lim m(E_n).$$

Proof. By Theorem A.5, the set E is measurable. We consider two possible cases.

1. There is k such that $m(E_k) = \infty$. Because $E_k \subseteq E$, we have $m(E) = \infty$ (cf. Exercise A.3) and the result follows.
2. $m(E_k) < \infty$ for all $k \in \mathbb{N}$. Since

$$E_{k+1} = E_k \cup (E_{k+1} \setminus E_k), \qquad k \in \mathbb{N},$$

we have, by Theorem A.3,

$$m(E_{k+1} \setminus E_k) = m(E_{k+1}) - m(E_k).$$

Furthermore,

$$E = E_1 \cup (E_2 \setminus E_1) \cup \cdots \cup (E_{k+1} \setminus E_k) \cup \cdots$$

with pairwise disjoint sets on the right side. By Theorem A.4,

$$m(E) = m(E_1) + \sum_{i=1}^{\infty}[m(E_{k+1}) - m(E_k)].$$

The nth partial sum of the series on the right side is

$$m(E_1) + \sum_{k=1}^{n-1}[m(E_{k+1}) - m(E_k)] = m(E_n).$$

Therefore, $m(E) = \lim m(E_n)$.   □

Theorem A.7. *Let (E_n) be a sequence of measurable sets such that*

$$E_1 \supseteq E_2 \supseteq \cdots \supseteq E_n \supseteq \cdots .$$

If $m(E_1) < \infty$, then

$$m\left(\bigcap_{k=1}^{\infty} E_k\right) = \lim m(E_n).$$

Proof. Let $E = \bigcap_{k=1}^{\infty} E_k$. We have

$$E_n = E \cup \bigcup_{k=n}^{\infty}(E_k \setminus E_{k+1}).$$

By Theorem A.4,

$$m(E_n) = m(E) + \sum_{k=n}^{\infty} m(E_k \setminus E_{k+1}). \tag{A.1}$$

In particular,

$$m(E_1) = m(E) + \sum_{k=1}^{\infty} m(E_k \setminus E_{k+1}).$$

Because $m(E_1) < \infty$, the series on the right side converges, that is,

$$\lim \sum_{k=n}^{\infty} m(E_k \setminus E_{k+1}) = 0.$$

The result follows from (A.1).   □

A.2 Measurable Functions over Arbitrary Sets

It is not difficult to verify that the statements of Theorem 2.29 hold for arbitrary measurable sets which were introduced in Sect. A.1. Therefore we use the same definition of measurable functions as in Definition 2.6. The only difference is that the measurable sets under consideration are allowed to be unbounded. Almost all statements in Sects. 2.7 and 2.8 are immediately seen to hold for this extended class of measurable functions (cf. Exercise A.8). The only exception is Egorov's Theorem 2.34 as evidenced by the following counterexample.

Let (f_n) be a sequence of functions on $\mathbb{R}$ defined by

$$f_n(x) = \begin{cases} 0, & \text{if } x \leq n, \\ 1, & \text{if } x > n, \end{cases} \qquad \text{for all } x \in \mathbb{R} \text{ and } n \in \mathbb{N}.$$

The terms of this sequence are measurable functions and the sequence converges pointwise to the zero function on $\mathbb{R}$. Suppose that the statement of Theorem 2.34 holds for this sequence and let $\delta = 1$ in the theorem. According to this theorem, there is a measurable set E_1 such that $m(E_1) < 1$ and (f_n) converges uniformly to the zero function on the set $E = \mathbb{R} \setminus E_1$. Then, for $\varepsilon = 1$, there is N such that

$$|f_n(x) - 0| < 1, \qquad \text{for all } x \in E \text{ and } n \geq N.$$

In particular, $f_N(x) = 0$ on the set E. However, this is impossible because the set E is not bounded above (cf. Exercise A.9).

We conclude this section by proving a theorem which is an analog of Egorov's Theorem 2.34 for arbitrary measurable subsets of $\mathbb{R}$.

Theorem A.8. *Let E be an arbitrary measurable set of real numbers, and let (f_n) be a sequence of measurable functions on E converging pointwise to a function f. Then E can be written as the union*

$$E = A \cup \bigcup_{k=1}^{\infty} B_k,$$

where the sets $A, B_1, B_2 \ldots$ are measurable, $m(A) = 0$, and (f_n) converges uniformly on each of the sets $B_1, B_2 \ldots$.

Proof. First, we consider the case of a bounded set E. By Theorem 2.34, for every $n \in \mathbb{N}$ there is a measurable set E_n such that $m(E_n) < 1/n$ and (f_n) converges uniformly on $B_n = E \setminus E_n$. We have

$$m\Big(E \setminus \bigcup_{k=1}^{n} B_k\Big) = m\Big(\bigcap_{k=1}^{n} E_k\Big) \leq m(E_n) < \frac{1}{n}.$$

Let

$$A = E \setminus \bigcup_{k=1}^{\infty} B_k,$$

so

$$E = A \cup \bigcup_{k=1}^{\infty} B_k.$$

Because

$$A = E \setminus \bigcup_{k=1}^{\infty} B_k \subseteq E \setminus \bigcup_{k=1}^{n} B_k,$$

for every $n \in \mathbb{N}$, and

$$m\left(E \setminus \bigcup_{k=1}^{n} B_k\right) < \frac{1}{n},$$

we have $m(A) = 0$, and the result follows.

Now suppose that E is unbounded. We can represent E as the union of pairwise disjoint bounded sets

$$E = \{0\} \cup \bigcup_{k=1}^{\infty} \left(E \cap ([-k, -k+1) \cup (k-1, k])\right).$$

By applying the result of the theorem for each bounded set in the above countable union, we obtain the desired representation.   □

A.3 Integration over Arbitrary Sets

As in Chap. 3, we begin by defining the integral of a nonnegative measurable function on a measurable set E (bounded or not).

Definition A.3. *Let f be a nonnegative measurable function on an arbitrary measurable set E. We define*

$$\int_E f = \sup\left\{\int_{E \cap [-n,n]} f : n \in \mathbb{N}\right\}.$$

Each of the integrals on the right side is well defined though it may be equal to ∞ (cf. Definition 3.2). According to our conventions, we set $\int_E f = \infty$ if one of the integrals on the right side is infinite. If $\int_E f < \infty$, we say that the function f is integrable *over E.*

Inasmuch as

$$\int_{E \cap [-n,n]} f \leq \int_{E \cap [-m,m]} f, \qquad \text{for } n \leq m,$$

for a nonnegative function f, we can write

$$\int_E f = \lim \left(\int_{E \cap [-n,n]} f \right).$$

The following theorem provides for a partial justification for the above definition (cf. the definition of a measurable set in Sect. A.1 and the subsequent theorem there).

Theorem A.9. *For a nonnegative measurable function f on E we have*

$$\sup \left\{ \int_{E \cap A} f : A \text{ is a bounded measurable set} \right\} = \int_E f.$$

Proof. We denote

$$S = \sup \left\{ \int_{E \cap A} f : A \text{ is a bounded measurable set} \right\}.$$

If $\int_{E \cap A} f = \infty$ for some bounded set A, then $S = \infty$. Then $\int_E f = \infty$ because there is n such that $A \subseteq [-n, n]$.

Suppose that $\int_{E \cap A} f < \infty$ for all bounded sets A. Because for any $n \in \mathbb{N}$, the interval $[-n, n]$ is a bounded set, we have $S \geq \int_E f$. On the other hand, for any bounded set A there is n such that $A \subseteq [-n, n]$. Hence, $S \leq \int_E f$, and the result follows. $\square$

For a function f of arbitrary sign, we use its positive and negative parts, f^+ and f^-, to define the integral $\int_E f$ (cf. Sect. 3.5).

Definition A.4. *Let f be a measurable function on an arbitrary measurable set E. We define*

$$\int_E f = \int_E f^+ - \int_E f^-,$$

provided that at least one of the integrals on the right side is finite. Otherwise, the integral $\int_E f$ is undefined. If $\int_E f$ exists and is finite, then the function f is said to be integrable *over E.*

Thus a function f is integrable on an arbitrary measurable set if and only if both nonnegative functions f^+ and f^- are integrable.

Theorem A.10. *If f is an integrable function over a set E, then*

$$\lim \int_{E \cap [-n,n]} f = \int_E f.$$

Proof. Inasmuch as f is integrable, the integrals $\int_E f^+$ and $\int_E f^-$ exist and are finite. Hence we have

$$\lim \int_{E \cap [-n,n]} f^+ = \int_E f^+ \quad \text{and} \quad \lim \int_{E \cap [-n,n]} f^- = \int_E f^-.$$

Therefore,

$$\lim \int_{E \cap [-n,n]} f = \lim \left(\int_{E \cap [-n,n]} f^+ - \int_{E \cap [-n,n]} f^- \right)$$

$$= \lim \int_{E \cap [-n,n]} f^+ - \lim \int_{E \cap [-n,n]} f^-$$

$$= \int_E f^+ - \int_E f^- = \int_E f,$$

and the result follows.   □

However, the existence of the limit $\lim \int_{E \cap [-n,n]} f$ does not imply Lebesgue's integrability of the function f as the following two examples demonstrate.

Example A.2. Let $f(x) = x$ on $\mathbb{R}$. Then

$$\int_{\mathbb{R}} f^+ = \int_{\mathbb{R}} \max\{x, 0\} = \infty \quad \text{and} \quad \int_{\mathbb{R}} f^- = \int_{\mathbb{R}} \max\{-x, 0\} = \infty.$$

Hence, the function x is not Lebesgue integrable over $\mathbb{R}$. However, $\int_{-n}^n f = 0$ for all $n \in \mathbb{N}$, so the limit $\lim \int_{-n}^n f$ exists and equals zero.

The limit $\lim \int_{-n}^n f$ may exist even for a bounded function over $\mathbb{R}$ which is not Lebesgue integrable as the next example shows.

Example A.3. We define

$$f(x) = \begin{cases} \dfrac{\sin x}{x}, & \text{if } x \neq 0, \\ 1, & \text{for } x = 0, \end{cases} \qquad x \in \mathbb{R}.$$

It can be shown that the limit of the sequence $(\int_{-n}^n f)$ exists (and equals π), but both Lebesgue integrals $\int_{\mathbb{R}} f^+$ and $\int_{\mathbb{R}} f^-$ are infinite (cf. Exercise A.11), so f is not Lebesgue integrable.

Many properties of the integral over an arbitrary measurable domain can be easily established as consequences of properties already established in the case of bounded domains (cf. Exercises A.10–A.15). However, some arguments are subtle. As an example, we give a proof of Fatou's Lemma (cf. Theorem 3.16).

Theorem A.11. (Fatou's Lemma) *Let* $(f_1, \ldots, f_k, \ldots)$ *be a sequence of non-negative measurable functions converging pointwise to a function* f *a.e. on* E. *Then*

$$\int_E f \leq \liminf \int_E f_k.$$

Proof. As in the proof of Theorem 3.16, we may assume that convergence takes place over the entire set E.

For any given $m, n \in \mathbb{N}$ we have

$$\inf \left\{ \int_{E \cap [-n,n]} f_k : k \geq m \right\} \leq \int_{E \cap [-n,n]} f_p, \qquad \text{for all } p \geq m.$$

It follows that

$$\lim_{n \to \infty} \inf \left\{ \int_{E \cap [-n,n]} f_k : k \geq m \right\} \leq \int_E f_p, \quad \text{for all } p \geq m. \qquad (A.2)$$

By Theorem 3.16,

$$\int_{E \cap [-n,n]} f \leq \lim_{m \to \infty} \left(\inf \left\{ \int_{E \cap [-n,n]} f_k : k \geq m \right\} \right).$$

By taking the limits as $n \to \infty$ on both sides, we obtain

$$\int_E f \leq \lim_{n \to \infty} \lim_{m \to \infty} \left(\inf \left\{ \int_{E \cap [-n,n]} f_k : k \geq m \right\} \right)$$

$$= \lim_{m \to \infty} \lim_{n \to \infty} \left(\inf \left\{ \int_{E \cap [-n,n]} f_k : k \geq m \right\} \right)$$

$$\leq \liminf \int_E f_k.$$

Here, the two repeated limits are equal by Exercise A.16, and the last inequality follows from (A.2). $\qquad \square$

The next two theorems can be proven by repeating verbatim the arguments used in the proofs of Theorems 3.17 and 3.25. The proofs are left to the reader as exercises (cf. Exercise A.17).

Theorem A.12 (The Monotone Convergence Theorem). *Let* (f_n) *be an increasing sequence of nonnegative measurable functions on* E. *If* (f_n) *converges pointwise to* f *a.e. on* E, *then*

$$\lim \int_E f_n = \int_E f.$$

Theorem A.13. (The Dominated Convergence Theorem) *Let (f_n) be a sequence of measurable functions on E. Suppose that there is an integrable function g on E that dominates (f_n) on E in the sense that*

$$|f_n| \leq g \quad on \ E \ for \ all \ n.$$

If (f_n) converges pointwise to f a.e. on E, then

$$\lim \int_E f_n = \int_E f.$$

Notes

The classes of measurable sets and integrable functions introduced in the Appendix are the same as obtained by more conventional methods. Of course, we cannot prove it here.

Theorem A.8 is known as Lusin's version of Egorov's Theorem (Theorem 2.34) for unbounded domains. Egorov's Theorem also holds in the following form for unbounded sets:

Theorem A.14. (Egorov's Theorem) *Let (f_n) be a sequence of measurable functions on a set E of finite measure that converges pointwise a.e. on E to a function f. Then for each $\delta > 0$, there is a measurable set $E_\delta \subseteq E$ such that $m(E_\delta) < \delta$ and (f_n) converges uniformly to f on $E \setminus E_\delta$.*

The following theorem is an important result which is also due to Lusin.

Theorem A.15. (Lusin's Theorem) *Let f be a measurable function on a set E. Then for each $\varepsilon > 0$, there is a continuous function g on $\mathbb{R}$ and a closed set $F \subseteq E$ for which*

$$f = g \ on \ F \ and \ m(E \setminus F) < \varepsilon.$$

The function f in Example A.3 has an improper Riemann integral

$$\int_{-\infty}^{\infty} \frac{\sin x}{x} \, dx = \pi.$$

On the other hand, this function is not Lebesgue integrable over the set of reals $\mathbb{R}$ [cf. Apo74, Exercise 10.9]. In this connection, see Exercise A.12 below.

As we observed in Chap. 3, Riemann integrable functions over a bounded interval form a proper subset of the set of Lebesgue integrable functions over the same interval. The situation in the case of unbounded intervals is different—the corresponding classes are incomparable in this case. In other words, there are functions which are Riemann (improper) integrable, say, over $\mathbb{R}$, but not Lebesgue integrable over that set (cf. f from the previous paragraph). On the other hand, the Dirichlet function,

$$f(x) = \begin{cases} 1, & \text{if } x \in \mathbb{Q}, \\ 0, & \text{if } x \notin \mathbb{Q}, \end{cases} \qquad x \in \mathbb{R},$$

is not Riemann integrable over any bounded interval, whereas it is Lebesgue integrable with $\int_{\mathbb{R}} f = 0$.

Note that we exchanged order of limits in the proof of Theorem A.11 (cf. Exercise A.16). For a general result see Sect. 8.20 in [Apo74].

Exercises

A.1. (i) Let (a_n) and (b_n) be two increasing sequences of real numbers. Prove that

$$\lim(a_n + b_n) = \lim a_n + \lim b_n.$$

(ii) Extend the result of part (i) to finite sums of increasing sequences.

A.2. Let E be a measurable set of real numbers. Prove that the set $\complement E = \mathbb{R} \setminus E$ is also measurable.

A.3. Prove that if E and F are measurable sets and $E \subseteq F$, then $m(E) \le m(F)$.

A.4. Let E be a measurable set. Show that

(a) If $m(E) < \infty$ and $\varepsilon > 0$, then there exist an open set G and a bounded closed set F, such that $F \subseteq E \subseteq G$ and

$$m(G) - m(E) < \varepsilon, \quad m(E) - m(F) < \varepsilon.$$

(b) If $m(E) = \infty$, then for any $M > 0$, there exists a bounded closed set $F \subseteq E$ such that $m(F) > M$.

A.5. Prove that the image of a measurable set E under the translation $x \mapsto x + a$ is measurable with $m(E + a) = m(E)$.

A.6. Prove Theorem A.5.

A.7. Let E_1 and E_2 be measurable sets. Show that

(a) The set $E_1 \setminus E_2$ is measurable.
(b) The set $E_1 \triangle E_2$ is measurable.

A.8. Show that the statements of Theorems 2.30, 2.31, Corollary 2.1, and Theorems 2.32, 2.33 hold for measurable functions over arbitrary sets.

A.9. Let A be a subset of $\mathbb{R}$ such that $m(A) < \infty$. Show that

$$\sup(\mathbb{R} \setminus A) = \infty.$$

A.10. Show that a measurable function f is integrable over E if and only if the function $|f|$ is integrable over E and that

$$\left| \int_E f \right| \le \int_E |f|$$

in this case (cf. Theorem 3.20).

A.11. Let f be the function from Example A.3. Show that

(a) $\lim \int_{-n}^n f$ exists.
(b) $\int_{\mathbb{R}} f^+ = \int_{\mathbb{R}} f^- = \infty$.

(cf. [Apo74, Exercise 10.9]).

A.12. Let $f : [0, \infty) \to \mathbb{R}$ be Riemann integrable on every bounded subinterval of $[0, \infty)$. Prove that f is Lebesgue integrable over $[0, \infty)$ if and only if the limit (the improper Riemann integral)

$$\lim \int_0^n |f| = \int_0^\infty |f|$$

exists. Show also that in this case

$$(L) \int_0^\infty f = (R) \int_0^\infty f.$$

A.13. Establish the linearity and monotonicity properties of the integral over arbitrary measurable sets (cf. Theorems 3.23 and 3.24).

A.14. Let f be an integrable function over the finite union $E = \bigcup_{k=1}^n E_k$ of pairwise disjoint measurable sets. Show that

$$\int_E f = \sum_{k=1}^n \int_{E_k} f.$$

A.15. Let E be a set of measure zero and f be a function on E. Show that f is measurable with $\int_E f = 0$.

A.16. Let (a_{mn}) be a double sequence of nonnegative real numbers such that $a_{mn} \ge a_{pq}$ for all $m \ge p$, $n \ge q$. Show that

$$\lim_{m,n \to \infty} a_{mn} = \lim_{n \to \infty} \lim_{m \to \infty} a_{mn} = \lim_{m \to \infty} \lim_{n \to \infty} a_{mn}.$$

A.17. Prove Theorems A.12 and A.13.

References

[Apo74] Apostol, T.M.: Mathematical Analysis, 2nd edn. Addison-Wesley, Reading (1974)

[Aus65] Austin, D.: A geometric proof of the Lebesgue differentiation theorem. Proc. AMS **16**, 220–221 (1965)

[BS11] Bartle, R., Sherbert, D.: Introduction to Real Analysis, 4th edn. Wiley, New York (2011)

[Bot03] Botsko, M.W.: An elementary proof of Lebesgue's differentiation theorem. Am. Math. Mon. **110**, 834–838 (2003)

[Bou66] Bourbaki, N.: General Topology. Addison-Wesley, Reading (1966)

[Hal74] Halmos, P.: Naive Set Theory. Springer, New York (1974)

[Knu76] Knuth, D.E.: Problem E 2613. Am. Math. Mon. **83**, 656 (1976)

[Kre78] Kreyszig, E.: Introductory Functional Analysis with Applications. Wiley, New York (1978)

[Leb28] Lebesgue, H.: Leçons sur l'intégration et la recherche des fonctions primitives. Gauthier-Villars, Paris (1928)

[Leb66] Lebesgue, H.: Measure and the Integral. Holden-Day, San Francisco (1966)

[Nat55] Natanson, I.P.: Theory of Functions of a Real Variable. Frederick Ungar Publishing Co., New York (1955)

[RSN90] Riesz, F., Sz.-Nagy, B.: Functional Analysis. Dover, New York (1990)

[Tao09] Tao, T.: Analysis I. Hindustan Book Academy, New Delhi, India (2009)

S. Ovchinnikov, *Measure, Integral, Derivative: A Course on Lebesgue's Theory*,   143
Universitext, DOI 10.1007/978-1-4614-7196-7,
© Springer Science+Business Media New York 2013

Index

S. Ovchinnikov, *Measure, Integral, Derivative: A Course on Lebesgue's Theory*,   145
Universitext, DOI 10.1007/978-1-4614-7196-7,
© Springer Science+Business Media New York 2013